The Component Contribution

The Component Contribution

Engine progress through the specialist manufacturers

Alan Baker BSc(Eng), ACGI, FIMechE

Hutchinson Benham, London

Hutchinson Benham Ltd
3 Fitzroy Square, London w1p 6jd

An imprint of the Hutchinson Group

London Melbourne Sydney Auckland
Wellington Johannesburg and agencies
throughout the world

First published 1979
© Associated Engineering Limited 1979

Set in Monotype Times New Roman

Printed in Great Britain by The Anchor Press Ltd
and bound by Wm Brendon & Son Ltd
both of Tiptree, Essex

isbn 0 09 136290 3

Contents

Acknowledgements

This book owes much more than most to other people. Since a lot of the required information is undocumented, mere reading could not be adequate, so I had to visit, correspond with and telephone a large number of engineers – individuals as well as members of companies – to gather my material.

My approaches were met with almost unvarying helpfulness; vast numbers of man-hours were lavished on discussions and internal researches on my behalf, whether for facts or for illustrations. A lot more time went also on 'vetting' the drafts, to ensure that I had correctly interpreted what my helpers had told me.

Without any exaggeration, the book could never have been written without such unstinting collaboration, so I should like to express my appreciation to all those who made it possible; I hope they feel that the end-product justifies the effort.

The following is a list (I trust a complete one) of my benefactors, and I would like to proffer my especial thanks to Associated Engineering who not only agreed enthusiastically to be the book's co-sponsors with Hutchinson but also were always ready with encouragement and practical assistance.

A C Delco Division of General Motors Ltd
Aeroplane & Motor Aluminium Castings Ltd
A E Turbine Components Ltd
Associated Engineering Ltd
Mr Donald Bastow
Robert Bosch Ltd

Brico Engineering Ltd
British Aluminium Co. Ltd
British Internal Combustion Engine Manufacturers' Association
Champion Sparking Plug Co. Ltd
Coopers Filters Ltd
Mr E. L. Cornwell
Covrad Ltd
Crosland Filters Ltd
Cummins Engine Co. Ltd
Engineering Components Ltd
Farnborough Engineering Co. Ltd
Mr Charles Fisher
Garrett AiResearch Ltd
General Motors Ltd
GKN Castings Ltd
GKN Forgings Ltd
The Glacier Metal Co. Ltd
Hepworth & Grandage Ltd
Holset Engineering Co. Ltd
Mr P. P. Love
Lucas Aerospace Ltd
Lucas Bryce Ltd
Lucas CAV Ltd
Lucas Industries Ltd
Mazda Car Imports (GB) Ltd
Noël Penny Gas Turbines Ltd
Renold Ltd
Ricardo & Co. (Engineers) Ltd
Serck Heat Transfer Ltd
Mr L. J. K. Setright
Mr Maurice A. Smith
Smiths Industries Ltd (Aviation Division)
Smiths Industries Ltd (Motor Accessory Sales & Service Division)
SU Fuel Systems (a British Leyland company)
Tecalemit Ltd
Turner & Newall Ltd
Uniroyal Ltd
Vandervell Products Ltd
Wellworthy Ltd
Western Thomson Controls Ltd

Zenith Carburetter Co. Ltd
and the librarians of the Institute of Marine Engineers, the
Institution of Mechanical Engineers, the National Motor
Museum, the Science Museum, and the Royal Aeronautical Society.

Foreword — A brief history of the internal combustion engine

There are no 'ivory towers' in technology, especially in engineering. The engineer who wants to advance dare not shut himself away, because the time must come when he cannot do it all himself – when he has to seek help from the specialists. So it has been and still is in the case of the internal combustion engine, and quite simply the purpose of this book is to underline the vital role played in the development of the IC engine by the specialist manufacturers of components and ancillary equipment. All too often they have been the unsung heroes of the struggle for progress, collecting any blame for failure but given little credit for success.

In terms of man's own history on this planet, that of the internal combustion engine may prove to be quite a brief one, though it will certainly not be unimportant. It all began early in the second half of the nineteenth century, with the invention of the gas engine as a rival to the steam engine, itself something over 100 years old by then. Thanks initially to the brains and hard work of pioneers such as Jean Etienne Lenoir, Beau de Rochas, Nikolaus Otto, Gottlieb Daimler, Dugald Clerk, Ackroyd Stuart, Rudolf Diesel and others, the IC engine made more rapid progress than did the steam engine during its corresponding development period. That rate of advance has been maintained, too, owing to the enormous strides made during the present century in design technique, metallurgy, petroleum technology (embracing both fuels and lubricants) and the study of combustion phenomena.

The large majority of internal combustion engines gain their energy from burning hydrocarbon 'fossil fuels' derived from crude petroleum, the world's resources of which are clearly not inexhaustible. Since the demand for these fuels has soared so steeply since the Second World War, conservation has already become a very real need. Even with strict economies, and allowing for the discovery of new major oilfields, the global reserves of petroleum look unlikely to last far into the twenty-first century. Consequently, unless alternative fuels can be produced in sufficient quantities from sources that are constantly being replenished by nature, the end of the IC engine's road could already be in sight.

As to the achievements of this type of 'prime mover', one is not exaggerating to say that, in a few decades, it has vastly altered the way of life of most of the world, primarily through the medium of transport-ation. Although the first IC engines were in fact used for industrial purposes, the march forward really began with their application to road vehicles. Initially these were of the passenger-carrying type, with four, three or even two wheels. Quite early in the twentieth century, though, came horseless vehicles for carrying goods and raw materials – originally over short distances only, since at that time reliability and durability could certainly not be guaranteed. Also near the start of this century, in 1904 to be precise, the first officially successful powered flight was made, using an internal combustion engine. Soon, flying and motor racing became popular recreations of the better-off; although automotive and aircraft power units continued to have much that was basically in common, they developed upon divergent lines because of their different operating conditions.

The 1914–18 war was a hot-house for the progress of the IC engine. In a few short years, aircraft versions developed from the relatively primitive – capable of little more than puttering a lightweight aeroplane across the Channel – to quite an advanced state at which power outputs of several hundred horsepower were being achieved, with a fair measure of reliability, for a big variety of military aircraft ranging from small fighters to multi-engine bombers and, in the case of Germany, bomb-carrying airships. Substantial progress was made on the ground too, and the hitherto impracticable became commonplace – motorcycles for dispatch riders, staff cars for senior officers, transports for war materials of all kinds, and even 'engines of war' themselves, namely armoured cars and the tanks whose relentless advance caused so much terror during their first appearance in action.

When peace returned in 1918, many of the lessons of war were applied elsewhere. Owing to the advance in the technology of the IC engine, mechanical transport flourished. Cars, motorcycles, buses, trucks and taxis proliferated on the roads while, in the air, commercial flying for passengers, goods and mail became a reality. At much the same time, in the countryside, the advent of the tractor began to revolutionize agriculture by vastly increasing the amount of work that could be accomplished in a day. The fields of marine and industrial power benefited equally from this progress, and locomotives with internal combustion engines instead of the traditional steam variety began to appear on certain countries' railways following some pioneering activities in pre-war days. Because of Britain's concentration, as a major coal-producing nation, on the steam locomotive, less was done here in this area than in a number of other lands. In fact, it was not until the 1930s that industrial-type IC engines were modified for locomotive applications, and then only for shunting duties.

Until the early 1920s most IC engines were of the gasoline (petrol) type with spark ignition, except in the marine and industrial fields where the heavy compression-ignition engine had already established itself. The latter variety has nearly as long a history as the former, and Ackroyd Stuart took out a patent on it in 1890, shortly before Dr Diesel, whose name has, however, become generally accepted as a generic term for this operating system. Because of major problems associated mainly with fuel injection and combustion, the compression-ignition engine did not progress so rapidly as the gasoline type before the First World War, but thereafter it made up for lost time, especially in the vehicle, locomotive and lighter marine areas. In these applications its higher thermal efficiency – and hence better fuel consumption – was of greater importance than the inferior power-to-weight ratio that results from the heavier construction necessary to withstand the higher peak pressures occurring in the cylinders.

Undoubtedly, the next major impetus to IC engine progress was provided by the 1939–45 world war, and the most spectacular advances were made in aviation. During the war, for example, aircraft-engine outputs leapt from around 1000 h.p. to 2000 and even 3000 h.p. in the continuing search for higher performance. These advances in the realms of orthodoxy, though, were overshadowed by the advent of a very different type of prime mover – the gas turbine. All motion in the gas turbine is purely rotary, so it has a basic advantage in terms of smoothness and internal loadings over the piston engine, in which reciprocating motion has to be converted into rotary. Once again, the principle of the gas turbine was not new (and the steam turbine had been known for centuries), but these power units had become practicable through improvements in materials, component design and manufacturing techniques.

Initially the gas turbine was adopted for aircraft (in 'jet' form) because, for a given power, it could be made lighter and with a smaller frontal area than could a piston engine; both these factors have of course a direct influence on the performance of an aircraft. Since the war years, though, the turbine's field has become much wider, primarily owing to its adaptation to driving an output shaft instead of merely exerting thrust through the expansion of the burning gases.

Nowadays the gas turbine has not only ousted the piston engine for aviation purposes in all but light aircraft but it is also quite widely used in marine and industrial applications – the latter particularly in the generation of electricity. It has been tried too in both cars and commercial vehicles, but scale effects and inferior response characteristics due to inertia have so far militated against its success in the first-named category. On the other hand, some knowledgeable vehicle engineers believe that the turbine may soon become the favoured type of power plant for long-distance coaches and heavy trucks.

Piston engines and gas turbines developed side by side during the early post-war period. In the case of the former, considerable advances began to be made in the specific power output (horsepower per unit of piston displacement) of automotive-type diesel engines through the use of turbo-superchargers driven by the otherwise wasted energy in the exhaust gases. Yet again, the principle was far from new, having been used for some years on heavy marine and industrial power units, but turbocharging had not really become generally recognized until applied to certain aircraft and tank engines during the war.

Then, in 1954, came the next milestone: Dr Felix Wankel, a German engineer, announced his rotary-piston engine which eliminated the reciprocating motion of the conventional piston layout. This ingenious design, which is believed to have been invented around 1922, promised smoother running and reduced bulk and weight, was quickly taken up in Germany by NSU for car applications, and at the time of writing is being produced under licence by several other companies also. The Wankel engine's early promise has been fulfilled in some respects, but in its basic form it has one or two inherent disadvantages, notably its relatively low thermal efficiency. Consequently it seems unlikely to predominate, although it will probably find a permanent place in the power-unit field for duties where the benefits outweigh the disabilities.

Within the last few years two factors have begun to have a profound effect on engine design, particularly in the vehicle context. One is atmospheric pollution and the other is the already mentioned rapid exhaustion of the world's reserves of crude petroleum. Both topics have been widely discussed at all levels, and this is hardly the place to embark on an inevitably long discourse of marginal relevance. Suffice it to say, therefore, that we have a conflict: some of the methods now being adopted to reduce the toxic exhaust emissions from vehicles have a significantly adverse effect on fuel consumption. Undoubtedly, more will be done in the future to reduce pollution by improving combustion characteristics, and it is conceivable that cleaner-burning fuels than petroleum derivatives will eventually be produced in sufficient quantities.

After rushing through this very potted history of the internal combustion engine, let us now stand to one side and see how the specialist component manufacturers came into being, and how their role has developed through the years. If a manufacturer is the only maker of a certain type of machinery in the country, and he wants a few items that are not already in production, it is probably cheapest and quickest to make them himself. Suppose then that his output increases enormously – what does he do? Rather than install a complete new manufacturing facility, he might well prefer to subcontract to a specialist who, if involved with products of a generally similar nature, already has the resources and the know-how.

On the other hand, maybe other manufacturers latch on to our first one's ideas and go into competition with machinery of the same type. There is then a strong likelihood that all these products would incorporate components of a similar kind and made in a similar way, though not necessarily identical. At once we have the opportunity for the specialist company to convince the others that it can do better for them collectively than they can do individually. In due course, pressure can be gently applied to the various rivals to modify their designs to accept standard components; these, being produced in much larger quantities than the original different items, should be significantly cheaper – an argument that appealed as much 100 years ago as it does today.

The next stage, as an industry expands and progresses, is for other specialists to appear in any particular area. This brings us to the situation of inter-specialist competition, with each one trying to obtain a bigger share of the available business than his fellows. In general, success here necessitates either a better product for the same cost or a cheaper one for the same performance – the ultimate of better *and* cheaper is rarely achieved! It follows that the rival specialists, in their endeavours to stay ahead (or catch up) have to set up their own research organizations, with interests in both the product itself and methods of making it as cheaply as possible. The resulting technological acceleration is always to the benefit of the customer firms and hence to the eventual purchasers of what they manufacture.

During the period since the Second World War, amalgamation into multi-company groups has been the industrial tendency. It applies as much in the sphere of the specialist engine-component maker as elsewhere, but the degree of selectivity appears to be greater in that it is usual for the major groups to be built up of companies having complementary interests. This situation has the major advantage of facilitating an integrated and therefore effective central research establishment. Although the 'fewer and bigger' pattern applies quite widely in many engine-producing countries, there are still large numbers of independent suppliers of all sizes. Also, some countries do a substantial export trade in components and accessories, selling them to areas where there is no indigenous industry or one that is less advanced. Thanks to her long tradition of engine manufacture, Great Britain is one of the bigger exporters in this field, particularly on the automotive side.

To conclude this introduction, some indication should be given of the range of products embraced by the term 'specialist components and accessories'. A comprehensive list would be almost impossible to compile and very tedious to read; by the same token, the succeeding chapters will cover only the more significant areas. First, clearly, come the structural elements of the engine – the crankcase, cylinder block and head or other items which house or carry the mechanical parts and assemblies and withstand the major operating stresses. Then we have the energy-transmitting side – crankshaft, connecting rods and pistons in the case of reciprocating engines, rotors in gas turbines and Wankel units, and of course the bearings in which they run. Other purely mechanical components include such things as valves and springs, the camshafts operating them and the chains, belts or gears that drive the camshafts.

No engine will run unless the fuel is fed into it in the appropriate quantity, and this involves carburettors or various types of injection equipment, according to the particular power unit. Gasoline engines require spark-ignition of the mixture, embracing the actual sparking plug and a means of providing the high-voltage spark at the right time. Again, engines have to be lubricated, cooled and started when required. This is no mean list, and there is no doubt that, if the specialist supply industry had not grown up and accepted its technological responsibilities, the internal combustion engine would today be markedly less far advanced than it is.

1 Automotive reciprocating power units

Introduction

GASOLINE ENGINES

The story begins with the gas engine (and here gas means gas – it is not the transatlantic contraction for gasoline) and the first such engine to go into production and service was that of Lenoir in 1860. In effect it was a simple adaptation of the steam engine, in which steam under pressure was replaced by the combustion of a mixture of town gas and air. This mixture entered the cylinder under atmospheric pressure and was then burnt without having been compressed, so power units of this type are known as 'atmospheric' engines. It is recorded that Lenoir's gas engine was tried in a road vehicle, which carried a large gas container, but the experiment was not a success.

Because of the atmospheric operation, Lenoir's engine had a very low power output in relation to its gas consumption. In 1862, though, de Rochas proposed the compression of the gas/air mixture in what was in effect the first four-stroke operating cycle. Two years later Otto in Germany started making gas engines, and he soon saw the possibilities of this four-stroke principle, though he did not make it fully practicable until 1876. Meanwhile, the first engine using a hydrocarbon fuel – naphtha vapour – had been built in the USA by George Brayton. It was of course an atmospheric engine and its only fundamental difference from Lenoir's was in the fuel employed.

Because by then Otto had obtained good patent protection on the four-stroke cycle, a number of engineers became interested in the two-stroke principle as a viable alternative. In the late 1870s two British pioneers – James Robson and Dugald Clerk – took out patents on designs for gas-burning two-strokes; the former incorporated under-piston scavenging of the residual burnt gases, and the latter a scavenge pump. Within a year or so, through the patents of two other Britons, Fielding and Day, these systems were joined by 'uniflow' (or through-scavenge) and loop-scavenge layouts, both of which are still in common use today.

Compression engines burning liquid fuels – kerosene (paraffin) and gasoline – instead of gas began to be evolved around 1880. And here we come to the inception of the motor vehicle, because in 1883 Daimler in Germany made his original four-stroke petrol engine, and in 1886 one was fitted to a bicycle which therefore became the archetypal motor-cycle. The following year Daimler built his first car, with an improved version of the same engine, and the transport revolution was under way!

Other engineers, some of whom – such as Benz – had been working on somewhat similar lines to Daimler, quickly became involved in the automotive field. The early engines were slow-running and by present-day standards produced very little power. However, progress was rapid, as is witnessed by the fact that the Daimler and Benz engines of about 1890 had a maximum speed of 600–700 revolutions per minute whereas the de Dion Bouton of 1895 could run up to 1500 rev./min. (If the reader is more familiar with the contraction r.p.m., I hope he will forgive me – I prefer to follow the British Standard!) Multiple cylinders came into use, in the search for smoother running and increased power, and many other ideas were tried (and some inevitably discarded) by those pioneers of motoring in the golden days before the First World War.

As was mentioned in the Foreword, racing (of both cars and motor-cycles) made substantial contributions to engine development – particu-larly in the pre-1914 era – because it stressed components and accessories to a higher degree than was possible in normal road use. Large numbers of companies became involved in the automotive field during this period, as manufacturers of vehicles, engines or essential bits and pieces. Inevitably, many of these firms have fallen by the wayside in the inter-vening years, but some of the early giants – such as Rolls-Royce and Ford, each of whom made a great contribution in its own way – are still with us today.

In terms of the vehicle gasoline engine, the period between the wars was to a large extent one of consolidation rather than innovation. Nevertheless, some real pioneering still went on, a noteworthy case in point being Herbert Austin's introduction of the first practicable small car – the original Baby Austin. This slackening of progress is probably not surprising when one considers the economic state of Germany, following the 1914–18 war, and the depressions in Britain and the USA. Since the Second World War, in contrast, the rate of development has been substantial, not least because of the greater degree of international competition.

Many of the developments have of course been concerned with achieving higher power and torque outputs, and with raising standards of reliability/durability. Very valuable progress has been made by the component and accessory firms in such respects as bearings, pistons and rings, carburettors, petrol injection, ignition equipment (through the adoption of electronics) and valve materials, for example. Aluminium has come into wider use in the search for lower weight and higher efficiency, principally for cylinder head and block castings in place of iron but also instead of copper and brass for radiator construction.

These and other advances are covered in the succeeding sections of this chapter. They have contributed largely to making today's spark-ignition automotive engines so much better in virtually every respect than those of the pre-1939 era. Sometimes our modern units have been criticized as being less economical of fuel than their predecessors, but in this context one must bear in mind the higher average speeds and greater traffic densities. In addition, the increasing emphasis on conservation nowadays is causing most engine manufacturers to strive harder for good fuel consumption, and these efforts are already bringing creditable results.

The four-stroke cycle continues to predominate in vehicle gasoline engines, the use of two-stroke power units being virtually confined to the motorcycle field. This is primarily because, in its basic 'three-port' form with petroil lubrication, the two-stroke's mechanical simplicity is offset by relatively low power or mid-range torque, and the environmental disadvantage of a smoky and smelly exhaust. Motorcycle two-strokes have been developed to a remarkably high level but it should be appreciated that not only do such engines have a correspondingly heavy fuel consumption but the characteristics suited to an enthusiast's machine of high power/weight ratio are very different from those for an ordinary car.

DIESEL ENGINES

Although Dr Rudolf Diesel, a German, has given his name generically to the compression-ignition engine, as stated earlier, it is widely considered that a Yorkshireman, Herbert Ackroyd Stuart, was working on the CI principle slightly earlier than Diesel. Certainly Ackroyd Stuart took out a British patent in 1890, by which time Diesel had not got beyond presenting a paper on the subject; he applied for a German patent shortly afterwards and for his first British patent in 1892. Their approaches differed, though, in that Stuart employed a pre-combustion chamber (which presented cold-starting difficulties) and Diesel an open chamber. Another difference was that Stuart had an airless injection system for his fuel whereas air-blast injection was used by Diesel. Engines of both types were in commercial production by the beginning of the twentieth century, but mainly for industrial duties; however, there are records of a road-vehicle application before 1890 and a locomotive one in 1902.

Steady progress was made in compression-ignition engines up to the First World War. The early power units were all of the four-stroke type, but the two-stroke principle was utilized shortly after the turn of the century, following on from the work of Robson and Clerk. Airless injection gradually took over from the hitherto more favoured air-blast system between 1910 and 1920, and pressure-charging (particularly as a means of improving two-stroke scavenging) came into use during the same period; it is worthy of mention that Dr Diesel envisaged the pressure-charging of a four-stroke engine as far back as 1896.

In the open-chamber layout as invented by Diesel, the fuel is injected directly into the cylinder, and so such engines are often referred to as being of the direct-injection type. Ackroyd Stuart's system, though, is known as indirect injection in that the fuel is injected into the pre-chamber where combustion is initiated. As time went on, it was found that, in general, indirect injection was better suited to smaller engines running at higher speeds, and various arrangements of this kind became popular for such power units, particularly on the European mainland. However, improvements to combustion-chamber forms and injection nozzles duly led to something of a revival of direct injection for other than the smallest-capacity diesels. In the vehicle sphere, the two basic layouts still exist side by side today.

Much of the pioneering work in the automotive use of the compression-ignition engine was done in Germany, Dr Diesel himself being convinced that this application would be an important one. Several experimental vehicle installations were made in 1910 and 1911 but were not highly regarded at the time, mainly because of the complication of the air-blast injection systems. Diesel and his collaborators realized that, for the automotive objective to be attained, airless injection would have to be more fully developed for use in place of the air-blast layout. It so happened that by then Robert Bosch, also a German, was perfecting the prototypes of his unitary plunger-type injection pumps which will be considered in greater detail later in this chapter. The First World War broke out, however, and the development of the vehicle diesel had perforce to be dropped in favour of more urgent projects.

The high-speed diesel engine really came into being with the rapid growth of road transport after the war. Germany again led the way (partly because she was short of petrol), and, MAN, Mercedes and Benz were among the early producers of vehicle diesels, the latter two amalgamating after a few years. In 1928 the first British automotive engine appeared, built experimentally by AEC, and it was soon followed by several others, among them a production model from Gardner – still a famous name in diesels today. The Gardner was in fact derived from one of the company's established industrial power units, whereas the Leyland diesel introduced in 1932 was designed essentially for vehicle use.

By the early 1930s, high-speed diesel engines were being fitted to a considerable number of commercial and public-service vehicles in various countries, primarily because of their good fuel consumption. In general mechanical terms, of course, the development of the automotive diesel engine followed very similar lines to that of the gasoline engine. One man who perhaps did more than anyone else to improve the diesel during this inter-war period was the late Sir Harry Ricardo (then H. R. Ricardo) through his intensive investigation of combustion phenomena. He was responsible for several successful indirect-injection systems and did much valuable work in evaluating the influence of the injection equipment. Credit must go too to Britain's Perkins company, which made a major contribution to popularizing the small high-speed automotive diesel.

Nowadays, of course, the diesel engine is the norm (except to some extent in the USA) for commercial and military vehicles other than lightweights, also buses and coaches, taxis, agricultural tractors and earth-moving equipment. It has made some inroads into the private-car sector, particularly in those countries where it is favoured by fuel tax differentials. In a straight comparison with the gasoline engine, though, its operational economy is offset by higher first cost and weight, generally lower specific power output and less refined running. Because of the first cost, it cannot show operational economy in a car unless big annual mileages are covered. Two-stroke and four-stroke diesels are both produced, the latter having a substantial lead in the vehicle field. Nowadays the exhaust-driven turbo-supercharger (or turbocharger) is finding increasing favour on both two-strokes and four-strokes as a means of raising power output and torque by harnessing the otherwise wasted energy in the exhaust gases. Such improvements in efficiency have recently become highly significant in the light of our greater awareness of the need to conserve the world's natural resources.

When one comes to consider the contribution of the specialist component and accessory manufacturers to the advancement of the vehicle diesel engine, there are clearly areas of departure from gasoline-engine practice – the other side of the technological coin from the parallel development mentioned earlier. Major equipment differences of course result from compression-ignition on the one hand and spark-ignition on the other: where the gasoline engine has a carburettor (usually) and a high-tension ignition system, the diesel has a fuel-ignition arrangement, nowadays comprising a high-pressure pump and nozzles. The diesel has to be controlled on the amount of fuel delivered to its cylinders, since it will not of course function if the air supply is throttled as in a carburettor. Other differences result also from variations in the working conditions provided by the two principles of operation.

In the diesel engine the maximum combustion pressure attained and the rate of pressure rise during burning are both substantially higher than in the gasoline engine. As a result the diesel has to be more robustly constructed, and its bearings, pistons and other components have to be capable of withstanding the heavier loadings. Temperatures too can be higher locally in diesel power units, so better-quality materials may have to be used, or special design features incorporated to prevent failures through overheating. These are the main reasons for the diesel's higher weight and cost, and the pressure situation accounts for its tendency to rougher running. In spite of these snags, though, its continued automotive popularity is assured by that significant fuel-consumption advantage, which stems from its fundamentally more efficient operating cycle.

Pistons, rings and cylinder liners

GASOLINE ENGINES

For many years the recipe for automotive gasoline engines was a cast-iron piston with cast-iron rings, running in a cast-iron cylinder bore. The combination gave no problems of compatibility or thermal expansion, and its wearing qualities were good at that time; also, since crankshaft rotational speeds were low, the relatively heavy piston did not give rise to excessive reciprocating inertia forces. From a very early stage the manufacture of pistons and rings was regarded as the province of the specialist companies, and even by the early 1920s a considerable amount of research work had been done by some of these on the production of satisfactory components to cope with increasingly arduous conditions.

At about that time, aluminium pistons were introduced for vehicle engines – not long after they had come into use on a number of aircraft engines – but they were not adopted on a large scale for several years. In the case of aero-engines the change was made primarily to save overall weight, but the reason was rather different in the vehicle field: engine speeds were rising quite rapidly, and inertia forces were therefore becoming troublesome in terms of bearing loadings and vibration. The main difficulties facing the engine and piston makers in going over to the lighter metal were wear and differential expansion. Aluminium on cast iron is basically a less good wearing combination than cast iron on cast iron. In addition, aluminium has an inherently higher rate of expansion with rise of temperature than has iron; consequently, there was the risk of a piston seizing in its bore when hot unless the cold clearance between the two was relatively large, and this resulted in considerable mechanical noise – piston slap.

The seriousness of both these problems was reduced metallurgically by the development by the piston companies of improved aluminium alloys. It was found that those alloys containing a substantial proportion of silicon (up to 13 per cent) were considerably harder and more wear resistant than the earlier materials, and they also had a significantly lower coefficient of thermal expansion. However, aluminium–silicon alloys still expanded appreciably more than cast-iron bores, so the piston experts (such as Hepworth & Grandage and Brico in Britain, the Mahle company in Germany and A. L. Nelson in the USA) began to investigate more complex designs in which some control was exerted over the expansion.

One of the first of these new designs to achieve commercial success was Nelson's Bohnalite piston. This was actually patented as early as 1927 and its objective was to channel the expansion away from the major thrust axis to the non-thrust axis where large cold clearances would have little effect on noise. The aim was achieved by casting-in two steel plate-type struts, each of which straddled one of the gudgeon-pin bosses; when the casting cooled, its contraction was opposed by the struts, putting the surrounding metal under a tensile stress. Then, when the piston heated up in service, the temperature rise had to relieve that stress before any expansion could occur.

Other companies developed strutted pistons operating on the same principle, and it is an interesting commentary on the rightness of the thinking that there is continued interest in them today, particularly in Germany. Because of the expense of casting-in struts, various other approaches have been made to controlling expansion. They include 'thermal slots' (to provide mechanical isolation of part of the crown from the skirt and so to minimize the transfer of heat from the crown to the skirt), as in the well-known Hepolite W-Slot piston, also the slitting of the skirt as a means of giving it a degree of flexibility, and winding steel wire helically round grooves in the piston body. Even where no such devices are used it has long been the practice to taper-machine the piston profile; this technique enables the relatively cool-running bottom of the skirt to have small cold clearances without those at the much hotter crown becoming too tight at the maximum operating temperatures. Also, since a piston is not a symmetrical shape, it does not remain truly circular on heating; another line of progress has there-fore been to design forms that are slightly elliptical when cold but tend to attain circularity under their normal operating conditions.

Since the Second World War, the power outputs and speeds of vehicle gasoline engines, and hence the thermal and mechanical loadings on their pistons, have markedly increased. As a result, piston makers have had to become considerably more scientific in their design and development techniques. That they have satisfactorily met the demands made on them is due to their evolution of precise methods of determining temperatures and stresses, and to improvements in materials and heat-treatment methods.

The post-war trend towards shorter-stroke engines brought its own special problems. Pistons had to be made shorter as well, and one consequence of this was a reduction in the number of piston rings, as will be referred to again later; shorter skirts inevitably increased piston stability problems, making the control of running clearances more critical than before. Difficulties have had to be overcome, too, in connection with the so-called bowl-in-piston engines. Since the pistons of such engines incorporate the combustion chamber (the underside of the head usually being flat), they have imposed additional design constraints in respect of dimensional precision and heat flow from the chamber surface to the rings.

From the very early days, the gudgeon pin has been regarded as an intrinsic part of the piston assembly and so its manufacture has been the prerogative of the piston maker. A surprising amount of research has gone into this simple-looking but heavily loaded component, to endow it with the necessary stiffness and fatigue resistance. Improved design, materials, production techniques, heat-treatment and finish have all contributed to the gudgeon pin's ability to withstand ever more stringent conditions. A noteworthy advance in this area was made by Hepworth & Grandage in the early 1960s: they evolved a technique of hardening the bores of the pins, with such marked benefit to the fatigue properties that it became possible to reduce the section or to employ a lower grade of steel, either change resulting in a worthwhile cost reduction. Another advance here was the adoption of cold extrusion for gudgeon-pin manufacture; it enabled the pins to be produced from small billets of steel, with minimal waste of material and good metallurgical structure.

Piston rings too are an essential part of the piston assembly, since they are responsible not only for forming a gas-tight seal with the bore but also for preventing excessive lubricating oil passing upward from the crankcase to be burnt in the combustion chamber. Cast iron has always been the most favoured material for piston rings, because of its good wear characteristics, but a problem encountered at the beginning of engine history was to get the iron rings to seal efficiently right round their circumference. This difficulty arose from the fact that the rings had to be 'gapped', not continuous, in order to cope with thermal expansion and manufacturing variations.

During the early years, various methods were tried of achieving good ring shape and providing the requisite tension to impart a uniform load round the periphery. It is worth recording that one of the most effective processes was developed in 1917 (in fact for the Bentley BR2 aero-engine) at South Coast Garages in Lymington, Hampshire; from this small works, then impressed into war service, grew the world-famous Wellworthy company. Their process, patented by the company's manager John Howlett and his foreman William Grey, was in fact based on an earlier Lanchester principle of non-circularity for piston rings. Howlett and Grey internally hammered a circular-cast ring in a die having an eccentric or cam shape; the stroke of the hammer was controlled to give maximum impact opposite the gap and less towards the ends, the blows being equally spaced. This hammering made the ring uniformly springy on its release from the die, and it became truly circular again when closed in the cylinder.

Although the germ of this idea is attributed to Dr Lanchester, the hammering of rings to alter their characteristics can actually be traced back as far as 1838, in steam-engine practice. The Howlett/Grey technique was one application of the process, another (which became more widely used) being to use a circular die but to obtain the desired 'free shape' by varying the spacing and/or strength of the hammer blows around the internal edge of the ring.

As production quantities of piston rings mounted, the individual hammering became unduly time consuming. From the 1930s it was therefore progressively phased-out in favour of a heat-forming process which had the additional advantage of increasing the strength of the ring. The heat-forming technique, though still used by some companies, has been superseded in other factories by the still more economical form-casting method in which, as the term indicates, the as-cast shape is carefully worked out so as to give the required distribution of spring. Most ring manufacturers today have adopted form-machining, a practice that involves the latest machine-shop technology.

The first piston rings were produced by parting them off from sand-cast pots or sleeves. In the mid 1930s the alternative of 'tree-casting' (in which from 50 to 120 rings are cast simultaneously yet discretely in the form of a stack or tree) was evolved in the USA, as a means of improving the uniformity of sand-cast rings, and the process was introduced into Britain by Hepworth & Grandage. However, pot-casting is still preferred by a number of major manufacturers, who usually employ the centrifugal-casting variation for the production of rings below about 6 inches (say 150 mm) diameter.

So far we have been considering piston rings generically, but in functional terms there are of course two distinct types – compression and oil-control rings. Advances in performance and changes of design in the automotive field have placed a heavy onus on the ring manufacturer in evolving satisfactory designs in both categories, and the more important developments will be discussed here. The axial width of compression rings, for example, has been progressively reduced to bring down the inertia as engine speeds have risen, and to give increased wall pressure. Before 1920, $\frac{3}{16} - \frac{1}{4}$ in. (about 4·8 – 6·3 mm) was usual, yet today car and motorcycle engines have rings as narrow as $\frac{5}{64}$ in. (about 2 mm) or even $\frac{1}{16}$ in. (say 1·6 mm); these narrower rings seal at least as effectively and are as durable as did their predecessors, thanks to improved design and production methods.

A major step forward in the early years after the Second World War was the chromium-plating of top compression rings to ensure adequate wear life in more powerful engines. This practice (pioneered during the war to combat abrasive desert conditions) became even more necessary when the number of rings was reduced, as mentioned earlier, since those remaining were more heavily loaded than before. The plating cut down the rate of bore wear as well as that of the ring itself. However, it had the disadvantage that its hardness resulted in slow bedding-in, and this in turn meant a big mileage before oil consumption came down to an acceptable figure. To overcome that difficulty, Hepworth & Grandage evolved their Cargraph treatment in 1959. This treatment consists of applying a small, carefully controlled quantity of abrasive to the working face of the ring after chromium-plating. The amount of abrasive has to be just enough to ensure rapid bedding-in wear for a very short period, yet insufficient to cause wear of the lower rings or the bore. After much painstaking experimental work the process was perfected and is now applied to a large number of plated rings for vehicle gasoline engines, particularly those for the replacement market.

A more recent development than chromium plating for top compression rings is spraying them with molybdenum. The process was first patented in Germany, and shortly afterwards in Britain and the USA; at present it is more widely used in the United States than elsewhere. It has the benefits of quicker bedding-in and freedom from the scuffing to which plated rings are sometimes liable in difficult conditions, but does not give quite such good wear resistance. As a halfway-house between standard rings and the plated or sprayed variety, in terms of cost and life, alloy-iron rings produced by sintering have lately gone into production by some piston-ring manufacturers, notably Brico Engineering, and are finding increasing favour. In some applications they have even been found superior to plated rings, certainly where there is any risk of scuffing.

Oil-control rings, which serve to limit the amount of oil getting up the cylinder bore, are usually very different in design from compression rings, though the bottom compression ring sometimes has a bevelled or stepped profile which gives it a marked controlling action. Looking back into history, special oil-control rings of any sort did not come into use on vehicle engines until the early 1920s, and the first ones were in fact of the stepped type. The slotted channel-section ring (which collects the oil in the groove between its two lands and returns it through further slots in the piston) did not appear until the late twenties. Since then the ring makers have produced numerous variations on the double-edged theme, some of them more elaborate than effective.

Many of these 'exotic' oil-control rings have been (and still are) of the steel multi-piece construction that originated in the USA in the late 1930s. To be really effective, such rings require expander springs to maintain them in proper contact with the bore, and various types of spring have been evolved for this purpose. Some of them have been of approximately polygonal form (originating in the mid-1920s for use with cast-iron rings), and so apply their pressure at a relatively small number of points; the more satisfactory designs, however, apply an evenly distributed pressure right round the ring. Interestingly, for all the time and trouble that has gone into the development of these multi-piece rings, there is still a preference in some countries – notably Britain and Germany – for the relatively simple cast-iron oil-control ring, which will work very satisfactorily provided that the cylinder bore remains round when the engine is hot.

Until the recent advent of aluminium cylinder blocks for quantity-produced cars, relatively few vehicle gasoline engines had renewable cylinder liners. During the 1930s, however, when bore wear was more rapid than it is today, one could have a worn block bored out and 'dry' cast-iron liners pressed in. Most such liners were simple sleeves, and some examples were chromium plated for minimum wear. The disadvantage here was the poor oil-retention characteristics of the plated surface, and the plating of the top ring instead duly proved to be a better and much cheaper solution. Both dry and wet (water-surrounded) liners are used on the new generation of aluminium engines in cars, depending on the designer's preference, so the specialists – often the same firms who make pistons and rings – have plenty of variety. They

Opposite: Some indication of the large number of possible sections for piston rings is given by this illustration, drawn from data provided by Associated Engineering Group companies

Plain compression

Bevelled scraper

Extra-duty oil-control

Tapered on periphery

Napier scraper

Narrow-land
Super Drain oil-control

land↕ Tapered on periphery,
with land

Double-bevelled scraper

Drilled and grooved
oil-control

Barrel-faced
chromium-plated compression

Stepped scraper

Double-bevelled
and slotted oil-control

Plain, internally stepped

Twin-segment
chromium-plated stepped scraper

Single-segment
oil-control with expander

Internally bevelled

Twin-segment stepped scraper

Twin-segment
oil-control with expander

Ridge Dodger

Slotted oil-control

Twin-segment oil-control

Dykes L -type,
pressure backed

Super Drain oil-control

Apex oil-control

Oil-seal, plain

Super Drain oil-control
with coil spring

Duaflex oil-control

Laminated steel

Super Drain oil-control
with expander

Abutment expander/equaliser
oil-control

also supply the motorcycle industry, since for weight reasons it is common practice today for the engines of two-wheelers to have aluminium cylinder barrels or blocks.

DIESEL ENGINES

As already indicated, the history of the automotive diesel really began in the early years after the First World War. Although engines of this new generation were referred to as 'high-speed diesels', the term was only a relative one since crankshaft maximum speeds were (and still are) generally lower than those of comparable spark-ignition power units, on account of the necessarily higher piston weights. The very first of these vehicle diesels – all of which were four-strokes – had cast iron pistons, following the convention which at that time had been broken only for aircraft engines. However, it is worth noting that, although in their case inertia forces were less important, the builders of high-speed diesel engines adopted aluminium pistons well before they came into widespread use for 'commercial' gasoline engines – even before quite a lot of car-engine manufacturers had said goodbye to cast iron. This more enterprising outlook may have been due to the fact that the vehicle diesel was really a new type of engine, and therefore not bound in the same way by metallurgical tradition.

So far as the piston manufacturer is concerned, the automotive diesel has proved in general more of a headache than the spark-ignition engine. In spite of the development of high-quality low-expansion aluminium alloys, the elevated temperatures and pressures of the diesel cycle inevitably pose expansion and strength problems. They also increase the criticality of dissipating the heat from the piston crown to the cylinder wall by way of the ring belt, and the thermal situation is complicated in the case of direct-injection engines by the incorporation of a toroidal or other shape of bowl in the crown. As a result of these factors, the aluminium diesel piston has evolved into something rather different from its gasoline-engine equivalent: it is longer in relation to its diameter, has a larger complement of rings and embodies more metal in the crown and gudgeon-pin bosses.

Even before the 1939–45 war, the intensive development of the automotive diesel was leading to substantial increases in output and consequently to still higher loadings, both thermal and mechanical. These very arduous conditions resulted in, among other things, unduly rapid wear of the top ring groove, through the combination of hammering and heat. The piston makers duly met that contingency by introducing various forms of inserts to carry the top ring. These inserts were usually of austenitic cast iron, which has a considerably higher coefficient of thermal expansion than has normal cast iron – not much below that of the piston material; they were secured by either casting-in or metallurgical bonding. The latter is generally regarded as the more satisfactory since there is no possibility of the insert loosening in service.

First in the field of metallurgical bonding was the Al-Fin process which was patented in the USA in 1944. It is interesting that Wellworthy in Britain, who had been working on bonding techniques for several years, finally perfected their own process in 1945. When they came to patent it, however, they discovered that it had been narrowly antedated by the Al-Fin one which turned out to be identical though arrived at entirely independently. However, Wellworthy came to an amicable arrangement with the patentees and were granted an Al-Fin licence for manufacture in the United Kingdom for world-wide sales; they also were given the power to grant sub-licences throughout the world.

The preceding remarks refer to pistons for four-stroke engines only. Two-stroke diesels for vehicle use began to be developed seriously during the late 1930s but they did not come into service in significant numbers until after the Second World War. Cast iron has long been the preferred material for the pistons of such engines, partly because of the particularly severe thermal and mechanical conditions encountered, and partly because relatively heavy pistons actually reduce bearing loads since their inertia opposes the high gas pressure during the compression and firing part of the cycle. However, the specialist manufacturers have found it necessary in some instances (particularly in opposed-piston engines) to adopt two-piece construction in order to achieve the desired standards of strength and durability. A typical example has a heat-resistant crown of austenitic iron bolted to a skirt portion of normal iron; the top ring groove is machined in the crown to ensure that its wear rate is kept to a reasonable level.

The biggest automotive diesel engines today are those used in earth-moving and constructional equipment. These power units are often turbocharged to give a high output and, because of this and their large cylinder bores, pose especial difficulties in respect of heat dissipation from the piston crown and ring pack. However, the problems encountered and the solutions evolved are very similar to those within the marine, industrial and locomotive fields. They will therefore receive their due measure of attention later in the book.

In general terms, developments in piston rings for automotive diesels have been along similar lines to those for gasoline engines. However, it is worth recording that the taper-section or 'keystone' ring (introduced about 1936 for aero-engines, again by Wellworthy, as a cure for top-ring sticking) proved very satisfactory on diesels in the early 1950s. Although tapered rings are still encountered, their use has been made less necessary by the development of oils of higher detergency and by improvements in piston design: by deepening the top land, for instance, the temperature at the top ring groove can be substantially reduced. In contrast to gasoline-engine practice, the multi-piece steel oil-control ring is not greatly favoured for diesels, for which the spring-loaded conformable control ring, usually with chromium-plated lands for maximum durability, is becoming widely used.

The chromium plating of top rings has long proved to be advantageous for long life in diesel-engine duties. It is still the yardstick of wear resistance, although molybdenum coating has found a substantial following as an alternative in some applications. An interesting high-temperature coating process known as 'plasma spraying', developed by Associated Engineering's research laboratories in Britain during the 1960s, was in fact first used by one of the AE companies, Wellworthy again, in 1967 for applying chromium to vehicle diesel piston rings of the larger sizes. Soon afterwards the process was adopted by Hepworth & Grandage, also in the AE group, and is currently used by them for the chromium coating of rings for a number of engines in regular production. In this process, which has been introduced also in the USA, the coating material, in powder form, is fed by means of an inert carrier gas through the plasma, where the temperature developed is so high as to render the particles molten. An extremely strong bond is therefore achieved with the parent material, while the inert gas prevents oxidation during deposition and assists in the control of porosity.

To help the bedding-in of diesel piston rings having a wear-resistant coating, it has become common practice to give the rings a ground, lapped or 'phonographic' (micro-groove) finish on the periphery. An effective peripheral form for chromium-plated rings is that given by 'barrel-lapping' which is claimed to provide good line contact between ring and bore right from the start, thus facilitating bedding-in; barrel-lapping was introduced not many years ago but is now recommended by virtually all piston-ring makers if running-in difficulties arc likely to occur. Another innovation by the ring specialists, for engines liable to top-ring scuffing, is the copper plating of the periphery as a dry lubricant;

in many cases the copper is applied on top of the chromium coating.

Cylinder liners have for years been much more commonly used in automotive diesels than in gasoline engines, primarily because of the emphasis on long bore life in commercial operation. Until the mid 1960s quite a number of vehicle diesels were of the wet-liner type, in which the liners are in direct contact with the cooling water. However, because of the prevalence of 'cavitation erosion' (a phenomenon that will be discussed in more detail in Chapter 3), and rigidity and sealing difficulties, the wet liner has gradually been phased out in favour of the pressed-in or slip-fit dry liner which is now the norm.

Cast alloy iron has long been the standard material for both wet and dry liners. The specialist manufacturers of these components have devoted a lot of time and energy to perfecting centrifugal and other casting techniques to ensure a good and homogeneous structure in the interest of a low and even wear rate. They have also made a close study of the optimum bore finish in relation to the initial bedding-in of piston rings; in this connection it is noteworthy that the old 'mirror finish' has been found much less satisfactory than one having a small but carefully controlled degree of roughness, forming oil-retentive hollows in a plateau-type bearing surface.

In the early 1950s, before chromium-plated piston rings became established, attempts were made to popularize relatively thin steel dry liners to allow closer bore spacing than was possible with the iron variety. The bores were chromium plated for wear resistance. Although the principle of these steel liners seemed sound enough, they were not very successful because of their liability to distortion under thermal gradients and varying degrees of interference fit.

Bearings

Inevitably, the bearing technology of the early years of the internal combustion engine was borrowed directly from steam-engine practice. It is fair to say that, since then, the major advances in respect of plain bearings and materials have largely stemmed from the automotive and aircraft fields. In the case of aero-engines, the motivation has of course been the need for extreme reliability and a high power/weight ratio. The vehicle-engine situation has been different, though, since economics and large-scale production contend with relatively high loadings and variable operating conditions.

T.C.C.—B

When the IC engine was born, by far the most widely used bearing material was that named after an American engineer, Isaac Babbitt. 'Babbitt metal', evolved for railway locomotive journal bearings in 1839, was originally a tin–antimony–copper alloy but several variations have been introduced through the years, the most fundamental of which has been the use of lead as a base instead of tin. The accepted term for this family of alloys has long been 'white metal', though the original name is still used – especially in the USA.

White metal has two characteristics that make it particularly suitable for crankshaft bearing duties: it has a high resistance to seizure when in contact with iron or steel, and it is relatively soft. The benefit of the first property is self-evident, and the second enables a white-metal bearing to accept minor misalignment and to embed small solid particles (dirt in the lubricant or the products of wear) into its surface, thus protecting the journal from being scored. It is interesting that, in spite of its age, white metal is still the best bearing metal in purely frictional and conformability terms, but its low fatigue strength at elevated temperatures became a disadvantage as loadings increased with rising power outputs and speeds.

Bronzes (copper–tin alloys) have an even longer history than white metal as bearing materials but, though they have a higher load-carrying capacity, they are less 'kind' and more liable to seizure, making them less suitable for crankshaft duties. They were joined in about 1870 by the first copper-lead alloys, which were rather better in these respects but at that time did not come into widespread use. However, they crop up again later in this short history.

Some specialist companies were already manufacturing bearings or bearing materials when the IC engine came into being, and others were formed as the demand for these power units increased. One such British firm, which has long since gained a worldwide reputation, is The Glacier Metal Co., established in 1899 to make white-metal bearing alloys. The reason for selling metals rather than complete bearings in earlier days was the method of producing the bearings. Then the usual technique was to cast the bearing alloy directly into its housing, which might be the connecting rod or cap in the case of a big-end bearing, or a thick steel or bronze shell for a main bearing; the cast metal was subsequently machined and hand-scraped to fit the journal. Many engine builders had their own set-up for the casting process, though some did contract the work out to bearing companies; even so, they still had to do the final fitting to the individual shaft.

The limitations of this production technique began to be apparent by the late 1920s in the USA, where the rate of car-engine building was rising very rapidly. Not only had many man-hours to be devoted to hand-fitting each set of crankshaft main and big-end bearings but it was becoming increasingly difficult to find the necessary numbers of skilled fitters. Thus the way was paved for the biggest single step forward in bearing technology, and one that undoubtedly made the mass-produced car engine a practicable proposition.

This miracle-worker was the 'thin-wall bimetal bearing' and the patentees of the design and manufacture were the Cleveland Graphite Bronze Co. It comprised a pair of strong but relatively flexible strip-steel half-shells about 1·5 mm thick, thinly lined with the bearing metal. The halves were produced to extremely fine limits on dimensions and surface finish, and they fitted into housings of precisely the right diameter to give the required installed bore of the bearing. Equally tight limits were imposed on journal diameter and finish, with the combined result that original or replacement shells – straight from the manufacturer – could be put into an engine by anyone, without any skilled fitting whatsoever.

Legal evidence given during patent litigation (see later) discounts stories that bearings of this exact type were employed by first Ettore Bugatti and then W. O. Bentley well before the Cleveland invention. There is no doubt, though, that the breakthrough owed something to the work of an American engineer, Bill Klocke, who began investigating bimetal bearings as far back as 1915. In 1926 Klocke began an association with Cleveland Graphite Bronze Co., which for about ten years had been producing bushes (one-piece sleeve bearings) from monometal strip, mainly for the motor industry. These bushes were made by 'wrapping' a length of the strip around a former; subsequent insertion of the bush into its housing closed the small residual gap.

Klocke was responsible for several CGB patents after 1926, covering both monometal half-bearings and bimetal bushes. However, the fundamental inventive step of the revolutionary thin-walled bimetal bearing shells was detailed in a patent taken out in 1932 by two of Cleveland's own engineers – John Palm and Ben Hopkins. Later that year, Palm and George Salzman, also of CGB, filed their patent for making the bimetal strip for these bearings. In their process, the bearing surface, initially white metal, was cast on to the steel backing.

Apart from the enormous speeding of production that resulted from this new technique, and the reduced component costs, the motorist's problems were reduced in that thin-wall bearings required much less running-in than did the old type; this was because of their precision of manufacture and the high standard of surface finish. Although there was no longer any significant risk of bearing seizure through hard driving at an early stage, some bedding-down still took place, however. Consequently, even today, several thousand miles must elapse before bearing friction falls to a minimum, so on this account alone a new engine cannot be expected to give the full performance of which it is capable.

Quite soon after the granting of the relevant patents, Cleveland were supplying their thin-wall split bearings and bimetal wrapped bushes (which appeared simultaneously on the scene) to several US quantity producers of vehicle engines, and their activities were attracting considerable interest outside America. In Britain a new plain-bearing manufacturer, Vandervell Products, had begun operations in 1932 and, appreciating the rapid growth of the motor industry in Europe, soon obtained a production licence from Cleveland – the only one that firm ever granted in the UK, though a German company was subsequently licensed also.

Vandervell actually began to produce thin-wall bearings in 1935. Initially, the bimetal strip was imported in coils from the USA, and the half-bearings were formed from that strip. Within two years, though, output had soared sufficiently to warrant a major extension of the UK manufacturing set-up. The first phase of this expansion was to set up a casting line for applying the white metal; during the proving stages for this line, tinned-steel strip was imported from the States. Finally, with the installation of a tinning plant, the way was clear for the all-British production of the bimetal strip. It is an interesting coincidence that the first suppliers of steel strip to the completed facility were the GKN group which Vandervell were to join some thirty years later.

Towards the end of the 1930s, rising power outputs and speeds were beginning to reveal the load-carrying inadequacies of the conventional thin-walled white-metal bearing. It had been found that a significant improvement in fatigue strength, and therefore in bearing life, resulted from a reduction in the thickness of the bearing metal on the steel backing strip. In the late 1930s, therefore, Cleveland, followed by Vandervell and Glacier, introduced 'thin' white-metal bearings having a lining thickness reduced from the customary figure of about 0·010 in. (0·254 mm) to about 0·005 in. (0·127 mm). Cleveland duly registered the trademark 'MICRO babbitt' – a name that has since become generic.

Soon, though, even these thin bearings began to prove inadequate, so attention turned to the possibilities of copper–lead bearing alloys. As far back as the 1920s these alloys were beginning to be used for aircraft and some vehicle applications necessitating a higher load-carrying capacity than that of white metal. Initially the copper–lead bearings were solid or had thick-wall backing, but in 1930 Cleveland brought out thin-wall copper–lead and even stronger lead–bronze components (the latter containing a significant proportion of tin) made by casting the bearing metal on to steel backing strip in the manner already described. At the time there was very little demand for them but, as indicated, the scene changed dramatically in under ten years, compelling a number of vehicle diesel-engine builders to change from white metal to copper–lead and, later, lead–bronzes for their crank-shaft bearings – in particular those for the big ends; the diesel is of course more severe on its big-end bearings than is the gasoline engine owing to its higher peak cylinder pressures and 'slogging' type of duty.

By the late 1930s the vehicle industry's requirements for thin-walled bearings had grown rapidly, and applications had been found too on other and heavier types of engine. Since Cleveland's cast-on copper–lead and lead–bronze bearings had now gained considerable favour for severe duties, Vandervell began production of these components (under their licence) in 1939, initially from bimetal strip imported from the USA. Although the original Cleveland patents were still valid, other bearing manufacturers – including Glacier, Vandervell's close competitors in some areas of the bearing field in Britain – had meantime felt compelled to take up the improved design to maintain their place in the market. The consequent litigation on infringement continued until the expiry of the patents in the early 1950s.

Some of these rival companies endeavoured to overcome part of the patent difficulty by investigating alternative methods of forming the bimetal strip into bearing shells. Also, the well-kept confidentiality of Cleveland's sophisticated processes of strip and bearing manufacture caused resort to a sintering and rolling technique for uniting the bearing material to the steel backing; this method, first developed by General Motors in the USA, was adopted shortly after the war by Glacier for copper–lead (and then lead–bronze) bearings.

Reverting to the pre-war situation, experience soon showed that in arduous conditions the cast-on copper–lead and lead–bronze alloys, because of their relative hardness and intolerance of misalignment, could give rise to unduly rapid wear of the journals unless the latter were hardened, for which the 'nitriding' process was then the most suitable. Since the hardening was not a cheap process, the engine makers naturally avoided it unless it was necessary too for added strength. The bearing manufacturers' answer to this difficulty was a trimetal or three-layer bearing. Here again Cleveland Graphite Bronze were first on the thin-wall scene (Albion Motors had tried a lead coating on thick-wall copper–lead bearings in 1934 as a running-in agent); Cleveland cast tin-base white metal on top of the cast-on lead–bronze lining. Vandervell imported some of these bearings for test but they failed disastrously because of diffusion of the tin into the lead–bronze layer.

A barrier coating of nickel proved a partial solution, but meanwhile Vandervell had heard that Pratt & Whitney had successfully tried a lead–indium plated-on overlay on the silver master-rod bearings of their radial aero-engines. The British company therefore imported from Cleveland a supply of steel/lead-bronze strip and proceeded to develop the lead–indium coating process at their Acton factory. So good were the resulting bearings that they were adopted for the Napier Sabre aero-engine which went into production in 1941/2; subsequently they became the archetype for heavy duties in automotive diesels and even a number of gasoline engines also.

The lead–indium overlay (which required no barrier layer) not only countered the journal-wear tendency – thus for some applications enabling unhardened journals to be used where they were otherwise practicable – but it also resisted corrosion and any tendency to 'scuffing' during the early bedding-down stages. These bearings were so successful in the UK that Cleveland and others investigated alternative overlays, for both copper–lead and lead–bronze bearings; the objective was to obviate the need of indium which is an expensive metal. Lead–tin-antimony was the first of these alternatives and after it came lead–tin and lead–tin–copper which showed better wear resistance where lubrication was poor; however, they required a barrier layer, commonly of nickel, and could not quite match the fatigue resistance and softness of lead–indium which remains one of the best overlays.

In the early post-war years, even the avoidance of hardening in many instances could not disguise the fact that three-layer automotive bearings were quite a bit more costly than white-metal ones. By then, too, even a number of car gasoline engines had become uprated to such an extent that white-metal bearings were no longer adequate. Consequently, a lot of research was done on finding an alternative bearing material that would be better than white metal but cheaper than copper–lead, and aluminium was one of those to be investigated. Aluminium-base solid bearings with tin as an alloying element had been around since the 1930s but, owing to the mechanical strength required to retain such bearings in their housings, the tin content had to be restricted to about 6 per cent; consequently the bearing-surface properties were not very good. However, once thin-walled components became established – with their steel backing to provide strength and stiffness – it was no longer necessary to compromise on the bearing characteristics of the lining material.

Glacier looked hard at aluminium in the light of this improved situation. Then in the early 1950s, following some close collaboration with the Tin Research Institute and the Fulmer Research Institute, they introduced the first of an entirely new range of aluminium-base lining materials containing sufficient tin (20 per cent in this instance) to confer excellent bearing properties. The tin was distributed throughout the aluminium as a continuous network, so the material was given the name 'reticular tin–aluminium'. Because casting-on posed problems of bond strength and embrittlement, the alloy was attached to the steel backing strip by a Glacier-developed continuous cold-welding method.

Reticular tin–aluminium proved to be an admirable compromise material for any but the heaviest duties. Its strength – and therefore load-carrying capacity – was rather more than $2\frac{1}{2}$ times that of white metal and not much below that of sintered copper–lead. In addition it was regarded by Glacier as being closer to white metal than to copper–lead or lead–bronze in its compatibility with unhardened shafts, its conformability to misalignment and its ability to cope with dirt without causing scoring. Because of these favourable characteristics, reticular tin–aluminium quickly gained popularity; it is currently used in a high proportion of European car engines as well as in a number of vehicle diesels.

This is not the end of the aluminium bearing story, however. Glacier continued to search for a material to meet the requirements of the highly rated automotive diesels that were being developed. Their objective was an alloy as strong as the better lead–bronzes, yet with superior resistance to seizure, wear and corrosion. In 1973, in collaboration with Associated Engineering's central R & D operation, they came up with a promising answer – an aluminium–silicon alloy similar to that used for many pistons but specially processed to modify the metallurgical structure, in particular to improve the ductility. When given a lead–tin overlay (on a nickel barrier layer) and used on hardened crankpins, these aluminium–silicon bearings seem to be a very good compromise for heavily loaded big ends, and they are of course equally satisfactory for the rather less arduous main-bearing duties. These bearings went into production in 1976–7 for some British Leyland and Perkins turbocharged automotive diesels and are apparently proving very satisfactory.

Clearly, no heavy-duty bearing can operate efficiently and durably without good lubrication, and this is especially true of plain bearings. In fact, the bearing makers themselves are the first to admit the debt they owe to the oil companies for producing lubricants that have kept pace with ever-increasing loadings. This achievement has not, of course, been gained in a vacuum but as a result of close collaboration between the oil, bearing and engine companies. It is not sufficient merely for a good oil to be provided: the bearing must be designed with due awareness of the engine and lubrication-system characteristics, so as to make the best use of that oil.

This leads us to one of bearing technology's greatest post-war advances – the adoption of the digital computer as a design tool. The theory of applying computer techniques to engine bearing design was first studied during the late 1950s in the USA and Holland, but without any real end-product. Then in the early 1960s the Engineering Department of Nottingham University, in association with Glacier and others, initiated a detailed investigation in the UK. Glacier (soon to become part of the Associated Engineering group) was subsequently one of the first companies in the world to make use of the techniques developed. A quicker method of solution was soon formulated at Cornell University, in the USA, and with Glacier's collaboration was developed into a practical and low-cost technique, well suited to the needs of the engine designers. By then, Vandervell were deeply involved with computer work and they arrived at their own method of breaking down the original complex programme into simpler, more comprehensible forms.

Since those days, computer-aided bearing design has become very widely used, right across the spectrum of engine types. By making the design process more precise it reduces the amount of expensive development time required. Not only does it enable the basic performance criteria to be established from the engine company's parameters, but it

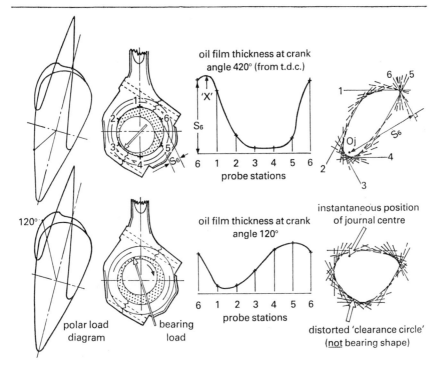

The measurement of bearing oil-film thickness is one of the modern predictive techniques used to speed development. These diagrams show typical dynamic distortion of a big-end bearing, determined by The Glacier Metal Co.

also allows the speedy assessment of the effects of alterations – for example, modifying the positioning of oil holes and grooves, adding or subtracting crankshaft balance weights or changing the viscosity of the lubricant. With the assistance of the computer, the design of plain bearings is now a much more exact science than it was twenty years ago.

Rolling-element (ball and roller) bearings have made a less significant contribution to vehicle-engine progress than they have in other fields. True, they were long the standard type for main and big-end duties in motorcycle engines, where they are still quite widely employed today, but one has the feeling that what started as expediency continued as habit. Bearings of this type have considerably less need of lubrication than plain bearings, and the oiling systems of earlier motorcycle engines (and of many two-strokes even today) could certainly be described as primitive.

It has been stated that the use of ball or roller bearings instead of plain ones reduces internal friction. However, this point is strongly argued today by the plain-bearing makers who maintain that only the static friction (i.e. when starting from rest) is significantly lower; in normal running, when the plain bearing has hydrodynamic lubrication, there appears to be little frictional difference. On the other hand, at relatively low speeds the load-carrying capacity of a rolling-element bearing is higher than that of a plain bearing of the same axial length, although at the expense of the radial space required for installation. Since motorcycle engineers have almost always adopted transverse crankshafts, they were probably influenced by being able to minimize the length of the crankshaft and hence the width of the crankcase.

Whatever the reasons for the selection of rolling-element bearings in the more distant past, it must be recorded that, since the late 1930s, the split-type plain bearing has made substantial inroads into the two-wheeler market. A factor contributing to the breakdown of tradition, especially so far as big-end bearings are concerned, has been the increase in the number of multi-cylinder engines. Producing a built-up crankshaft for these is appreciably more complex and costly than where there is only one cylinder, and the alternative of roller bearings with split races introduces other problems.

Before we leave the subject, reference should be made to 'dry' bearings – those that need no additional lubricant. For many years, porous oil-retaining bronze bushes have been used where loadings are moderate and speeds are relatively high – as, for example, the bearings for fans or dynamo armatures. If these bushes are impregnated with oil before installation, they will run for long periods without further lubrication, though their performance is always improved by an oily environment. However, applications do arise where oil lubrication is impracticable or intolerable, and self-lubricating graphited-metal bearings are one answer in these circumstances. Such bearings have been in existence since before the end of the nineteenth century, but the most substantial advances have been made since the end of the Second World War.

Powder-metallurgy methods (compacting and sintering) are used to produce these bearings which can contain up to about 14 per cent of graphite by weight – ample to confer permanent self-lubrication. One of the best-known ranges of graphited materials are the Deva Metals, first patented in Germany in 1948 and subsequently made in the UK, under licence from Deva Werke, by two Associated Engineering companies – initially by Universal Metallic Packing and later by Glacier. In 1975 AE acquired Deva Werke and the company was renamed Glacier GmbH-Deva Werke.

The first Deva Metals were bronze-based, but since the mid 1950s the range has been considerably extended and now additionally embraces lead–bronzes, tin–bronzes, iron and nickel; all Deva bushes were monometal until the 1960s, when the bimetal steel-backed variety was developed. The appeal of these graphited metal bearings for difficult duties is enhanced by their good mechanical properties, tolerance of dirt and insensitivity to unusually high temperatures. A noteworthy example of their application is in the primary gear transmission of British Leyland's transverse engines. The original input-gear bearings had to be positively lubricated from the crankshaft, but it proved difficult to keep the oil out of the clutch; Deva Metal bushes were therefore tried instead, with only oil-mist lubrication, and BL have used them there ever since.

The other major post-war contribution to dry-bearing technology was the discovery in the late 1940s of the very slippery polytetrafluoroethylene, or (thank goodness!) PTFE for short. Glacier was the world's first bearing company to develop a satisfactory method of incorporating PTFE in its products; this was achieved by impregnating a porous, sintered-bronze matrix with a PTFE–lead mixture. Following intensive testing, the Glacier DU range of thin-walled steel-backed bushes, made in this way and needing no lubrication at all, was introduced in 1958.

Within a few years other bearing manufacturers – including Vandervell and Railko in the UK and Du Pont and General Motors in the USA – were producing PTFE-based bushes in competition with Glacier. However, since the DU material remains probably the most widely used of this type, it is reasonable to assume that no one else has achieved a better balance of properties.

In spite of their special characteristics, and their successes in many machinery duties, PTFE-based bearings have so far found relatively few applications in the engine field. Those so far established include bearings for starter armatures, alternator rotors, water-pump impellers and throttle butterfly valves in carburettors. Other possibilities that have been investigated are the bearings of tensioner pulleys for overhead-camshaft driving belts, also camshaft and rocker bearings; the last two would not usually be running completely dry but tests have indicated that they would not require force-feed lubrication. The bearing makers have therefore helped the engine designer by reducing the number of constraints under which he has to work.

Carburettors

Since the spark-ignition engine inhales a mixture of air and fuel (not strictly true since there have been one or two examples of diesel-type injection of fuel directly into the cylinder), one of the first essentials is a device for metering both in the correct proportions. The most common means of doing this – or trying to do it – on early vehicle gasoline engines was the surface carburettor. In effect this was merely a box containing fuel; the air passed over the surface of the fuel on its way to the engine, collecting the evaporations as it went. The primary disadvantage of the surface carburettor was the wide variations in mixture strength due to corresponding changes in the evaporation rate. Such influences as the volatility of the gasoline (itself very variable in those days), the ambient temperature and even the roughness of the road all had their effect on the amount of fuel picked up by the air.

Something better was clearly needed, and that 'something' was the spray-type carburettor, with float control of the fuel level. In such instruments, of course, the fuel delivery from the jet into the airstream is governed basically by the difference between the pressure in the float chamber and that in the venturi where the jet outlet is situated. The spray-type carburettor appeared shortly before the end of the nineteenth century, and by the early 1900s it was already being made in some quantities by specialist firms such as Longuemare in France.

Although superior to the surface variety, these primitive spray-type devices left much to be desired. Because of their technical simplicity they were not able to compensate for the effect of engine speed and load variations on fuel delivery and therefore mixture strength. Consequently the throttle control had to be supplemented by an air control with which the driver had to juggle as he drove, endeavouring to match the air/fuel ratio to the immediate requirements of the engine. Since there was probably yet another lever for varying the ignition timing, many a motorist then must have longed for at least one more hand!

The specialist manufacturers got busy on these problems, and by 1907 the 'automatic' carburettor had appeared, albeit with a small effective speed range. Its automaticity, in terms of modifying the fuel flow to compensate – at least partially – for changes of speed and load, stemmed from the use of multiple jets instead of the single jet of the original instruments. In theory, the automatic carburettor obviated the need for frequent operation of the air control. A few years later the first practicable variable-jet carburettor came on the market. In this, the rate of fuel delivery was related to the engine load by connecting the accelerator control to a tapered needle within a fixed jet; opening of the 'throttle' raised the needle, thus increasing the annular gap between it and the jet, and allowing more gasoline to emerge. The variable-jet principle, with refinements, is still employed today in the majority of motorcycle carburettors.

An inherent feature of the conventional carburettor was that its fixed-size venturi or choke resulted in major variations in the velocity of the airflow with throttle opening, and therefore of the fuel velocity also – a major factor in the need for a compensation system. This problem caused G. H. Skinner, one of the pioneers of carburation, to investigate and patent in 1905 a very different approach – the constant-vacuum carburettor, which his S U company (still going strong today within the British Leyland empire) first put on the market before the

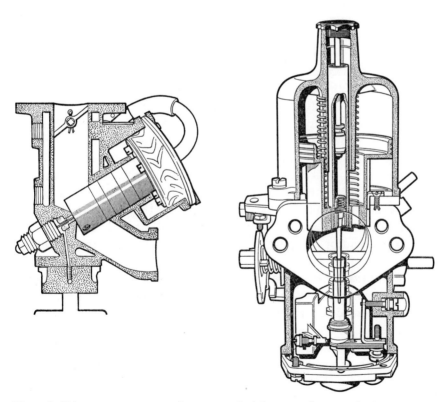

The early SU constant-vacuum carburettor on the left operated on exactly the same principle as does its present-day counterpart

First World War. The principle of such carburettors is that the choke area is varied with the engine demand so as to keep the depression in the choke constant; because of this constant depression, the air and petrol velocities remain constant also, their actual flow rates depending on the effective sizes of the choke and jet. Choke size is varied by a movable plunger subjected to induction depression, and jet size by a tapered needle attached to the plunger. In the original S U design, a flexible leather bellows was used to actuate the plunger, but later instruments had a metal piston instead. It is noteworthy that, in spite of the soundness of Skinner's principle, the S U was the only constant-vacuum carburettor in normal production until after the 1939– 45 world war, when it was joined by others of US and Japanese design. The most outstanding of these was the Stromberg C D instrument, in which the depression is applied to a flexible diaphragm instead of a piston – an interesting return to something near the first S U arrangement.

Reverting now to fixed-choke carburettors, their manufacturers introduced numerous improvements during the inter-war years, in the continuing search for better performance allied with lower fuel consumption and better cold starting. One such advance was the introduction of the accelerator pump in about 1924 in the USA. The normal carburettors of those days gave a hesitant response to sudden opening of the throttle, through temporary over-weakness, unless they were set unduly rich for steady running conditions. To overcome this, Schebler were the first to incorporate a small pump, operated from the throttle linkage, to squirt a little additional fuel into the choke on 'snap' acceleration. Not only did this scheme (nowadays a standard feature of fixed-choke carburettors) ensure a clean pick-up but it enabled leaner cruising-mixture settings to be used, with benefit to economy. Schebler were followed, also in the States, by Stewart-Warner and Stromberg, and the accelerator pump was then taken up in Europe during the early 1930s.

Other devices evolved by the carburettor makers during the 1920s included improved slow-running systems, 'emulsion-tube' arrangements that gave considerably weakened mixtures for part-throttle cruising, 'power jets' and special starter devices to replace the strangler valve (also sometimes called the 'choke', but this leads to confusion with the venturi which of course has the same alternative name). The power jet – another US idea that did not spread to Europe until the 1930s – was brought into action by the last stage of opening the throttle, to give a slightly over-rich mixture. This served the dual purpose of ensuring that full power was available and that internal temperatures were kept down, neither of which would apply if the mixture were on the weak side.

The various special cold-starting enrichment arrangements that appeared during this period embraced a secondary jet system and even an auxiliary starting carburettor – both embodied in the main instrument and brought into action by a separate control. Some layouts incorporated a means of automatically reducing the enrichment as the engine warmed-up. However, these systems tended to be complicated and not always reliable, so the practice more recently has been for carburettor manufacturers to revert to the strangler valve with either manual or automatic control, the latter being thermostatic by means of a bimetallic element. In both cases the old scheme of having separate hand throttle control for obtaining a fast warm-up idle has been superseded by an interconnection between the strangler and the throttle.

The so-called 'automatic choke' is in fact only semi-automatic since the driver has to set it initially by fully depressing the accelerator which then has to be prodded shortly after a cold start to reduce the idling speed. A much-superior approach to cold-starting is the Zenith FASD (fully automatic starting device) announced early in 1977 after a long gestation period. This is in effect a small auxiliary needle-type carburettor which really is what the name says, thanks to several features including thermostatic control of the mixture enrichment through a temperature-sensitive capsule immersed in the coolant and connected to the metering needle.

Since the Second World War further substantial progress has been made in carburettor design. Apart from the development of other constant-vacuum layouts, as already mentioned, the outstanding feature of the last two decades has probably been the arrival of the multi-barrel fixed-choke carburettor – a layout in fact pioneered by Zenith as far back as the 1920s. Most of these carburettors are of the twin-barrel type but four-barrel units have been developed in the USA by Rochester and others for big V8 engines.

There are two types of multi-barrel carburettor. Of these, the first has two identical chokes, fed from a common float chamber. The two throttles are operated in parallel by the accelerator-pedal linkage and the unit functions exactly as though it were two single-choke carburettors; it is, however, appreciably more compact. In some installations the two chokes feed a common inlet manifold, while in others each supplies one half of a duplex system. Such carburettors, two or even three of which are sometimes employed, can be regarded as merely a means of increasing top-end power output by improving the engine's breathing capacity.

The more subtle application of the twin-barrel principle is in the 'compound' carburettor. It provides a solution to the difficult problem of obtaining satisfactory metering at low speeds while retaining big enough orifices for good high-speed breathing. At idle and light load, only one choke (the primary) is in action, the throttle of the other being closed, so accurate metering is relatively easy. As the accelerator is further depressed, though, the secondary choke comes into action – either through a mechanical linkage or a vacuum-operated control – to give adequate air-passing capacity for higher loads. A refinement on the basic design is to have the primary choke of smaller diameter than the secondary one, to allow still better control of the idling and light-load mixture strength. In a four-barrel carburettor the principle is identical but there are two primary and two secondary chokes.

Before we leave carburettors, reference must be made to the additional difficulties raised by the present-day emphasis on the reduction of atmospheric pollution by exhaust emissions. Here, yet again, the carburettor and other ancillary equipment manufacturers have been heavily involved. For instance, in those countries such as the USA and Japan where the regulations are particularly stringent, carburettors providing specially close tolerances on fuel metering have to be fitted. Another line of attack is the provision of thermostatic control of the temperature of the incoming air; in other cases a 'gulp valve' is installed to let additional air into the inlet manifold on sudden shutting of the throttle, when the mixture is rich and might otherwise burn in the exhaust system rather than the cylinders. On sustained over-run, however, the mixture becomes progressively weaker, often beyond the limit for combustion, so heavy emissions of unburnt hydrocarbons can result. The latest SU carburettors therefore incorporate a small poppet valve in the throttle butterfly to limit the manifold depression and hence to avoid over-weakness. Yet another anti-pollution device – evolved by Ford for their own carburettors – is an idling system that gives particularly good atomization of the fuel; this system enables the idling mixture setting to be considerably weaker than usual, and permits the strangler to be dispensed with significantly earlier after a cold start.

The present-day ultimate in controlling the air/fuel ratio is the 'feedback' or closed-loop system which is applicable either to carburettors or to petrol injection. It has been evolved to enable full advantage to be taken of the latest exhaust-cleaning equipment – the 'three-way catalyst'

The wealth of sophisticated equipment used today in the development of engine components is evident from this view of an instrumented engine installed in a test cell. *Associated Engineering Ltd*

Sectioned Hepolite Pyrostrut piston for a car engine, showing
the cast-in steel inserts that control thermal expansion.
Hepworth & Grandage Ltd

Opposite: These groups of pistons, for gasoline engines
(above) and diesel engines (below) illustrate the wide range of
variables that has to be met by the specialist manufacturers.
Hepworth & Grandage Ltd and Wellworthy Ltd

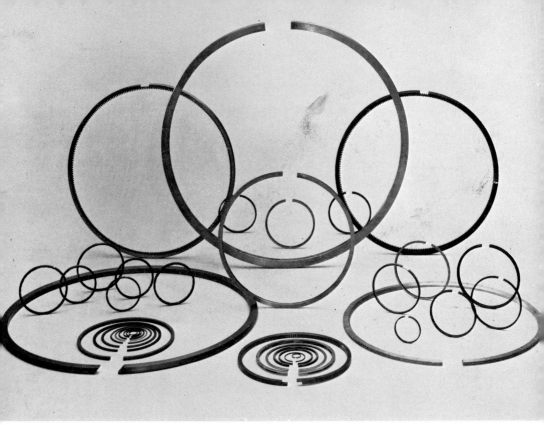

Though simple in appearance, piston rings involve a lot of advanced technology as well as being produced in a great variety of sizes and types. *Associated Engineering Ltd*

By means of this eight-channel telemetry linkage attached to the big-end, a direct read-out of piston and ring behaviour can be obtained in the laboratory while an engine is running. *Associated Engineering Developments Ltd*

A 'tree' of individually cast piston rings for vehicle engines, seen after removal from the sand mould. *Hepworth & Grandage Ltd*

For high-volume production, the machining of automotive thin-wall bearings is here being carried out on a linked line of machine tools. *The Glacier Metal Co. Ltd*

Opposite above: Piston rings for large automotive diesel power units being parted-off from the cast-iron 'pot' which represents the first stage in their manufacture. *Wellworthy Ltd*

Opposite below: A group of thin-wall bearings, having a copper–lead lining on a steel backing, for engines small and large. *The Glacier Metal Co. Ltd*

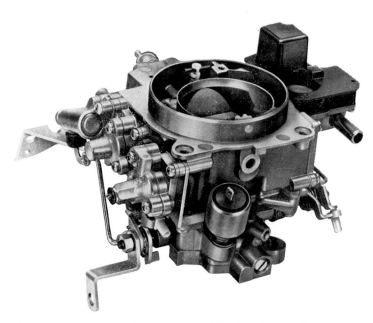

Present-day car carburettors, as typified by this downdraft fixed-choke instrument, have to be complex and precisely manufactured to ensure low fuel consumption and acceptable exhaust emissions. *Zenith Carburetter Co. Ltd*

Constant-vacuum car carburettors undergoing a series of acceptance tests in the factory after assembly. *Zenith Carburetter Co. Ltd*

Gasoline injection has supplanted carburettors on many high-performance and high-cost car engines. Here are the electrically driven pump and the metering unit of the Lucas Mark 2 system. *Lucas Industries Ltd*

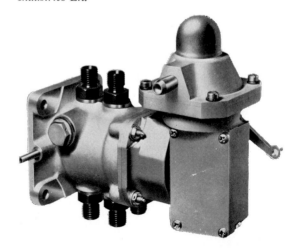

which deals with all three main pollutants in one pass, but only over a narrow band of air/fuel ratios straddling the chemically correct one.

In the feedback arrangement a sensor in the exhaust manifold detects the considerable changes of oxygen content that take place as the mixture strength changes through that narrow band. The sensor's signal is amplified and used to modulate either the fuel or the air delivery to correct excessive variations. At the time of writing, a control system of this kind had already been introduced by Volvo, in collaboration with Bosch, on a PI installation, while Zenith in Britain and Holley in the USA were perfecting carburettor applications.

Fuel-injection equipment

GASOLINE ENGINES

The injection of gasoline into the engine, as an alternative to carburation, is effectively a post-war development which to some extent stemmed from the military aero-engine field. Two of the main objectives of injection – improved volumetric efficiency (with consequently higher power output) and more even mixture distribution between the various cylinders – were the same as for aircraft power units. However, a third one, that of insensitivity to attitude, was less important in motor vehicles than in combat aircraft, though it was of some significance in racing because of the relatively high accelerations involved – both lateral and longitudinal. Such accelerations can of course affect the mixture strength of a carburettor engine by altering the fuel level relative to the jet system.

It should perhaps be explained why injection is potentially better than the carburettor in respect of volumetric efficiency and distribution. In the first place, the choke of the carburettor inevitably has a restrictive effect on airflow, and this is amplified by the necessity of some degree of preheating to obtain good 'atomization' of the fuel on its way to the engine. On the distribution side, a carburettor normally feeds into a manifold, and thence to the inlet ports, so some degree of asymmetry – and therefore of mixture bias – is almost inevitable in a multi-cylinder engine. Even in the case of high-performance power units, with multiple carburettors and one choke per cylinder, there are considerable practical balancing problems, and it is possible to get worse mixture distribution than with one carburettor and a well-designed manifold. In theory, an injection system overcomes both disadvantages: there is no need of a venturi in the inlet tract, and the metering arrangement should enable the same quantity of fuel to be delivered to every cylinder, though there is an unconventional layout here to which subsequent reference will be made. Unfortunately, the benefits are offset by the substantially greater cost of injection equipment; this aspect too is mentioned again later.

The gasoline can be injected either directly into the cylinder, as in diesel practice, or into the inlet tract upstream of the valve. In the latter case, the injectors can be sited either in the cylinder head or in the individual external branches of the induction system. Direct injection has so far found little favour, though it may yet come into its own through the increasing interest in stratified-charge combustion systems as means of reducing exhaust emissions and fuel consumption. There are numerous examples of both cylinder-head and external injectors, the latter having the advantage of the greater ease of 'ringing the changes' during the development stages. The actual injection can be either intermittent or continuous, but only the former, with precise timing, can really give the sought-for perfect fuel distribution. Regarding the earlier reference to the 'odd man out', there is a current system in which the induction tracts of each three cylinders of a six-cylinder engine are fed simultaneously – not in exact relationship to the operating cycle. Although this method works well enough in practice, and simplifies the equipment, it too cannot support any claim for perfect distribution.

Apart from racing, for which gasoline injection is always preferred today where the regulations permit, quite a number of car manufacturers nowadays offer injection on top-of-the-range models, primarily as a means of providing a higher power output. However, in one or two instances the objective is also better control of exhaust emissions: if the fuel metering is sufficiently precise, into the individual induction tracts, the result can be a leaner *average* mixture than is possible with a carburettor/manifold set-up where bias often occurs and the setting has to provide an over-rich mixture for some cylinders to avoid excessive weakness for the others. Conversely, there have been cases where, for markets having stringent anti-pollution requirements, an injection installation has been replaced by carburettors because of the former's inadequacy on the control side. It is a not uninteresting commentary on this aspect that in the USA – the birthplace of pollution-consciousness – carburettors are still preferred for the native products.

Relatively few specialist firms have become involved in injection systems for gasoline engines – among them, Lucas, Bosch and Kügel-fischer all have systems on current production cars, while others have a limited application for competition or high-performance conversion purposes. The manufacturers have had to work very closely with the engine makers on individual applications to match the characteristics of the system to the requirements of the engine. As the basic control parameter, either throttle angle, inlet-tract depression or mass airflow is used, and the actual delivery of the injection pump is controlled either mechanically or electronically. A few years ago it looked as though electronic control (covering also such additional operational factors as air and coolant temperatures, barometric pressure – or altitude – and sudden changes in inlet depression) would become standard practice. However, in 1973 the German Bosch company – who have as much experience on fuel injection as anyone in the world – reverted from electronic to mechanical control on their latest mass-airflow system, thus throwing the technology wide open again.

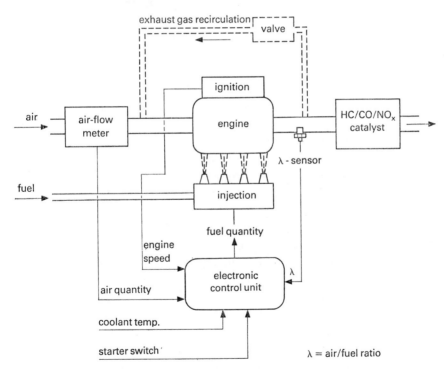

Diagrammatic representation of a Bosch automotive fuel-injection system with closed-loop or feedback control of the air/fuel ratio, in the interest of minimal exhaust emissions

There is no doubt that a lot of hard thinking, careful design and pains-taking development work have gone into bringing petrol-injection systems to their present level of refinement. However, their adoption in place of carburettors necessitates their showing significant perfor-mance advantages and these can be achieved only by very accurate manufacture, sophisticated control arrangements and careful matching to engine needs, all of which cost a great deal of money. Consequently, fuel injection for gasoline engines looks likely to remain the prerogative of the more expensive, higher-performance type of road car and the competition vehicle. As was once said by Charles Fisher, one of Britain's leading experts in this field, if injection had come on the scene first and then the carburettor had been invented, the latter would have been welcomed with open arms as the solution to so many problems!

DIESEL ENGINES

In terms of fuelling, the diesel is an entirely different animal from the gasoline engine. For the latter, the fuel is metered into the air in the correct proportion, by either a carburettor or an injection system, and the two enter the cylinder as a mixture. Speed and power are governed basically by the throttle, which varies the amount of mixture inhaled and is under the direct control of the operator. The fuel of a diesel engine, however, is delivered separately from and later than the air; since the latter is always supplied in excess of the theoretical require-ment, the speed and load have to be regulated by varying the quantity of fuel supplied. This function is performed, in the case of a vehicle engine, partly by the operator (through the accelerator or its equivalent) and partly by an automatic device called a governor.

In the introduction to this chapter, reference was made to the fact that early attempts to adapt the diesel for automotive duties were not very successful. This was because at that time – around 1910 – the injection and atomization of the fuel was generally by means of a blast of air at pressures as high as 900 to 1000 pounds per square inch (62·1 to 68·9 bars). Such pressures necessitated a sizeable air compressor and reservoir which were not readily accommodated on a motor vehicle.

After peace returned in 1918, Robert Bosch in Germany was able to resume work on his airless injection method which subsequently did so much to popularize the vehicle diesel. In his system, a multi-element in-line pumping unit delivered fuel to a series of spray nozzles, one in each cylinder head. His pump was of the reciprocating-plunger type and became known as a 'jerk pump' since the plungers (one per cylinder) were actuated sharply in turn by cams on a common shaft to force small volumes of fuel at high pressure along pipes to their respective nozzles.

Not only was the Bosch pump compact and mechanically sturdy, and able to operate consistently at relatively high speeds, but it had an ingenious and improved method (not a Bosch invention, incidentally) of varying the fuel delivery. Previously, several methods of doing this by altering the stroke of the pump had been tried but none had proved very satisfactory, particularly on the smaller sizes of engines. The new system consisted of keeping the stroke constant but having a variable 'spill point', at which fuel ceased to be delivered to the nozzle but instead was recirculated to the pump intake. To achieve this variation, the plungers were rotatable and embodied a circumferential groove having a helical upper edge; according to the angular position of the plunger, this edge uncovered a spill port in the cylinder higher up or farther down the stroke. All the plungers in the pump were rotated in unison by means of a longitudinal toothed rack engaging similar teeth on a 'control sleeve' on the lower part of each plunger.

Bosch decided that, for vehicle purposes, the interests of simplicity would best be served by the use of a governor that controlled only the idling and maximum speeds – the former for traffic stops and the latter to prevent engine damage through over-revving. Control of inter-mediate speeds in relation to the load was left to the driver's right foot. The governor, which was of Harzmann design and interposed between the control rack of the pump and the linkage to the accelerator pedal, was of the conventional centrifugal type; its bobweights were acted upon by two sets of springs – light ones for the idling-speed control and stronger ones for limiting the maximum speed.

As the compression-ignition engine began to be accepted for vehicle usage, other specialist manufacturers came on the scene in competition with Bosch. They all based their designs on the 'jerk' principle, and most of them evolved some sort of variation on the constant-stroke/variable-spill idea. Britain was not so quick off the mark here as some other European countries, but in 1933 CAV started to make pumps to the Bosch design. Subsequently, the British company embarked on its own development programme and today is one of the world's largest manufacturers in this field, in a time when production has become increasingly concentrated in the hands of a relatively small number of major organizations.

Various refinements have been introduced on in-line injection pumps through the years. Among them is automatic control of the injection timing in relation to engine speed; this type of control, analogous to that used for the contact-breaker of spark-ignition engines, retards the timing at low speeds as a means of reducing the tendency to the traditional 'diesel knock' and maintaining optimum timing through the speed range. Governors too have received a lot of attention from the specialist firms. The mechanical type is still widely used but has in many instances been elaborated into an 'all-speed' or torque-sensitive unit, having the final say in the pump delivery throughout the range of accelerator-pedal travel. Alongside these governors, though, other designs have appeared, notably the cheaper but less accurate pneumatic variety and the more sophisticated hydraulic types. In return for their higher precision, the latter are generally more costly, however, so the vehicle operator has to determine his priorities before specifying his equipment.

So far we have been considering only the in-line reciprocating-plunger type of injection pump. An interesting 1940s variation on this, and one that gained some popularity in the USA for a while because of its compactness, was an axial-piston design. Its pumping elements were grouped in a ring around the driveshaft axis, on which was an inclined swashplate operating the plungers.

Recently, an entirely different type – the distributor pump – has taken over from the in-line unit as the most widely used automotive equipment. As so often seems to be the case, the operating principle – a single pumping element feeding a distributional system to the nozzles – is far from new. It was first applied during the early 1940s by Hartford in the USA but achieved no commercial success at the time. Not until 1956 did the distributor pump really 'get off the ground', when CAV in

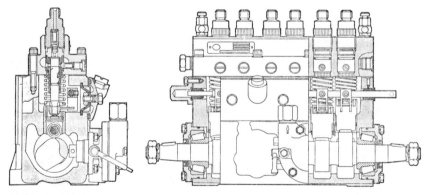

Rigid construction was a primary design requirement of the Lucas CAV Maximec in-line fuel-injection pump recently introduced for large automotive diesels

Britain took up the Hartford design, modified it to meet cis-Atlantic requirements and laid down a substantial production facility for what they christened their DPA unit.

In the automotive context, the distributor pump has several inherent and significant advantages over the in-line type. In the first place it is more compact (though little more so than the axial-piston layout just mentioned), simpler and cheaper, and provides uniformity of delivery to all cylinders. Because of its light reciprocating parts too, it can operate at higher speeds than can a comparable reciprocating-plunger pump; 12000 (6 × 2000) deliveries a minute are comfortably within the scope of present-day units. An additional appeal of the DPA pump when it first appeared was that, owing to its precise manufacture, it required the engine builder to make only one adjustment, namely setting the maximum delivery of fuel.

The DPA unit consisted essentially of a rotary pumping element within a stationary housing. In the element was a diametral bore containing two plungers which were actuated simultaneously at intervals by pairs of lobes on a cam-ring in the housing; there were as many pairs as the engine had cylinders. Fuel from the tank was fed by a transfer pump to a metering valve, the output of which was controlled by a mechanical or hydraulic governor.

From the valve, the fuel travelled through intermittently aligning drillways in the housing and rotor into the space between the plungers which it pushed apart ready for their next actuation by the cam-ring. As the lobes moved the plungers inward, another drillway in the rotor registered with one of the delivery ports of the pump. Maximum delivery was limited by mechanically restricting the amount by which the plungers could be separated by the incoming fuel, and automatic advance of the injection timing could be effected by rotating the cam ring by external means.

So well was the CAV distributor pump received by the vehicle industry that it clearly could not go long unchallenged by other injection-equipment manufacturers. Both Robert Bosch in Germany and American Bosch (for some time independent of the Stuttgart firm) were among those who introduced products of the same kind, the former as long ago as 1963. The various companies have all made significant improvements to their distributor pumps through the years, in respect of both performance and versatility. As a result, this type of equipment, originally intended for the smaller diesel because of its high-speed capability, is now used most of the way up the automotive size scale and has been adopted also for engines outside the vehicle field.

Although the pump is the most important and expensive part of the fuel-injection system of a diesel, efficient combustion depends to a great extent on the injector nozzles in the cylinder head. Here, too, much work has been carried out by the specialist manufacturing companies, and research organizations such as Ricardo, in close collaboration with the engine builders (in all diesel categories – not merely the automotive) on obtaining the most suitable fuel spray for each combustion-chamber layout. While much has now been learnt about the interrelation of engine and injector characteristics, each new power unit still requires a joint optimization programme. The best compromise has to be found between atomization and penetration – the former in the interest of easy ignition and the latter for thorough mixing with the air.

Where a jerk pump is used, there is normally no mechanical operation of the nozzle, injection being automatic as the fuel pressure rises. In early high-speed automotive engines equipped with this type of pump, the nozzles were of the 'open' variety, having merely a non-return valve to prevent any flow back to the pump owing to cylinder pressure. Such nozzles were simple to make and not liable to blockage, but they did not give precise timing of the beginning and end of injection and were sensitive to the bore and length of the supply pipe from the pump. 'Closed' nozzles were therefore soon introduced and came into general use. They are so called because the fuel-way to the nozzle tip is shut off by a valve until the delivery pressure has risen sufficiently to overcome the resistance of the valve's control spring; when the pressure falls again, the valve snaps back to its seat. The opening and closing characteristics can be adapted to suit the particular engine by altering the valve lift and the strength or preloading of the spring.

Injection equipment of the types described, comprising a simple remote pump unit and separate nozzles, has become standard on automotive diesels except for two well-known US ranges – Cummins and General Motors Detroit – whose makers fit systems of their own design featuring individual combined pump and nozzle assemblies. In theory, these two systems do not fall within the frame of reference of this book, since both are produced by the companies concerned, not by outside suppliers. However, the relevant divisions operate in exactly the same way as does a specialist manufacturer, so to exclude their equipment as being 'home made' would be illogical. It is relevant to explain here that although the diesel-building industry developed later in the USA than in Europe, Cummins – the pioneers across the Atlantic –

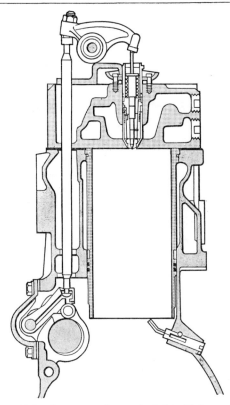

The Cummins PT diesel injection system has an individual injector for each
cylinder, actuated by a pushrod and rocker from an additional lobe on the camshaft

began manufacture before unit pumps became available there. They
therefore had no option but to develop their own equipment, and their
example was soon followed by General Motors when evolving their
engines for large-scale production.

In both these systems, which today are still basically the same as
when they were introduced in the 1930s, each plunger is actuated from
the engine camshaft by an auxiliary rocker. The main difference between
them lies in the method of varying the fuel delivery. Cummins do this
by having a governor-controlled low-pressure metering pump which
delivers the required quantity of fuel to the injectors. G M C, on the
other hand, have a constant-stroke/variable-spill internal system
similar to that originated by Bosch. Technically there seems to be little
to choose between the integrated and separate-pump designs, though
the smaller bulk of non-pumping nozzles makes them easier to accommodate on the engine.

Superchargers

GASOLINE ENGINES

The supercharging of vehicle gasoline engines, to increase their power output, has never been more than a minority activity, but it is of interest in the historical context because of the variety of proprietary equipment used. It should perhaps be explained briefly at this juncture that supercharging means pushing more air into the engine than could be achieved by atmospheric means. The greater weight of air is matched of course by a correspondingly increased delivery of fuel, so the energy input is higher and hence one can expect to get a greater power output. How much is actually gained depends on the power consumed (if any – see later) in driving the supercharger.

As has so often happened in the past, racing provided the incentive for the first efforts in supercharging, during the 1920s, since at that time the regulations did not impose an arbitary handicap on this means of getting more power. The trend spread into the sports-car field, resulting in such famous vehicles as the 'blower' Bentleys and supercharged Mercedes, Alfa Romeos and M Gs of the era extending from the late twenties well into the thirties. Most of those earlier blown cars had super-chargers of the Roots positive-displacement type. These were engine-driven, sometimes off the front end of the crankshaft, and consisted of two meshing twin-lobe (figure-eight-shape) rotors within a housing having the form of two 'siamesed' cylinders. Since the Roots blower does not compress its charge internally, the supercharging effect came from running it at a high enough speed for it to deliver more mixture than the engine would otherwise inhale. Though reasonably efficient at moderate speeds and boost pressures, Roots superchargers had a fluctuating delivery because of their configuration, and the gears that coupled the rotors were usually decidedly noisy.

In the search for a lower noise level and better low-speed boost, several specialist companies in Britain and on the European mainland brought out superchargers of the eccentric-vane design, another positive-displacement type but with the advantage over the Roots of having internal compression. All these eccentric-vane superchargers comprised a cylindrical rotor revolving within a cylindrical casing but on a parallel axis to give the desired eccentricity. In the simplest form the rotor had radial slots in which the flat vanes were a sliding fit. When the supercharger was revolving (usually by means of crankshaft-driven belts), centrifugal force kept the vanes in contact with the casing. This arrangement was not very satisfactory because of the friction between vane tips and casing, and between the vanes and their guides owing to the former's angularity to the casing over part of the swept path. One means of overcoming this objection was the 'guided-vane' arrangement developed pre-war by Zoller in Germany and then Centric in Britain, and continued post-war by Shorrock, also in Britain. In these superchargers the vanes were usually supported in trunnions, were always radial to the casing and were accurately constrained radially so that they did not rub on this.

During the 1930s two or three US car manufacturers offered supercharged models. The superchargers were not of the positive-displacement but the centrifugal type used on many aircraft engines; in these, air (or mixture) is drawn in axially and then flung out radially, by a rapidly revolving impeller, into a volute-shape collector leading into the induction tract of the engine. Because their impellers were mechanically driven and centrifugal force varies with the *square* of the angular velocity, these superchargers gave little boost at low engine speeds and so were theoretically the worst possible variety for normal road cars, for all their quietness and smooth air-delivery characteristics. However, they did improve mixture distribution at the lower speeds by their stirring-up effect.

Since the end of the Second World War there has been a small market for bolt-on superchargers (eccentric-vane) for the keen road driver who wants more performance without the disadvantages of normal 'tuning' techniques. On the other hand, interest in supercharging for racing engines fell off sharply post-war: in the motorcycle field, positive-displacement blowers (as used pre-war by BMW and DKW in Germany) were banned unless their swept volume was below that of the engine, and for cars an unfavourable supercharged/unsupercharged size differential was introduced. Almost the only racing organization of any significance to opt subsequently for a small supercharged engine was BRM, whose original V16 $1\frac{1}{2}$-litre Grand Prix unit – competing with $4\frac{1}{2}$-litre unblown ones – had a gear-driven aircraft-type two-stage centrifugal supercharger designed, built and developed by Rolls-Royce. The application was not unreasonable for a high-speed racing engine, but owing to many teething troubles the BRM did not develop its expected power output of well over 400 b.h.p. until shortly before the demise of the $1\frac{1}{2}/4\frac{1}{2}$-litre formula.

During the last few years the hitherto unpopular centrifugal super-charger has begun to make an impact in the realms of the gasoline engine – not in its original mechanically driven form but as part of an exhaust-driven turbine/supercharger (turbocharger) assembly which takes no power from the crankshaft. This is a development borrowed from the aircraft and diesel-engine fields, and will be discussed at greater length later. Suffice it to say that suitable well-tried proprietary units were already in existence but that some control problems have had to be solved to obtain adequate low-speed performance without over-boosting – and the consequent risk of engine damage through deton-ation – at higher speeds. Applicational techniques are now becoming fairly well understood, and turbocharged engines have achieved many successes in certain classes of racing. The turbocharger is beginning to establish itself for specialist road cars too, as a means of providing high performance (with docility, reasonable fuel economy and good exhaust-emission characteristics) from a relatively small engine. Judging it on an overall performance/cost basis, it could well extend its foothold.

DIESEL ENGINES

Before the Second World War very little work had been done on super-charging automotive diesel engines to increase their specific power output, partly because there was no real incentive in that direction beyond a progressive increase in running speeds. Subsequently, though, came a general trend towards improving efficiency so that vehicles could carry heavier loads or travel faster, or smaller engines could be used for a given duty, with benefit to initial and operating costs. It was therefore understandable that supercharging should soon have come under investigation.

Only two types of supercharger have found favour for vehicle diesels. One is the Roots or lobe type and the other is the exhaust-driven turbocharger; both have already been mentioned in connection with automotive gasoline engines, and the second will crop up again later. The Roots blower has been applied mainly to two-strokes, for scavenging as well as pressure-charging the cylinders. It was chosen because of its efficiency at the modest speeds and delivery pressures required, and only recently has its usually rather high noise level become a significant dis-advantage in the vehicle context. During the years, the type has been developed by the various makers in respect of such matters as the rotor-coupling gears (to minimize noise), rotor and casing form, materials, bearings and lubrication. Generally speaking, though, it can be regarded as an accessory – a package that is bought by the engine builder and bolted on to it to do a certain job.

In contrast, the turbocharger is really a component of the engine, since it is closely involved with the exhaust side as well as the inlet, and the installation has to be carefully matched to the engine's characteristics. This is the process that demands close collaboration between the makers of both the turbocharger and the engine if the full benefits are to be realized. Turbocharging can be used on two- and four-stroke power units but on the automotive front four-stroke applications have always been the more numerous – not least because of the preponderance of engines of this type!

The idea of driving a turbocharger by means of the exhaust – and thus making use of some of the otherwise wasted energy of the fuel – was conceived many years ago by a Swiss engineer, Alfred Büchi, of the Sulzer engine company. As far back as 1906 he took out a British patent on using the exhaust of an I C engine to drive a turbine, and then he thought of coupling the latter to an air compressor to form a supercharger that took virtually no power from the engine. Büchi began his experimental work on turbocharging four-stroke diesel engines in 1911, and by 1914 had established the basic viability of his system. Some research was carried out elsewhere during the First World War on applying the turbocharger to aero-engines; then in the early 1920s, during that era of rapid progress, turbocharging began to be adopted commercially for large marine and industrial diesels. However, as indicated at the start of this section, some twenty-five years were to elapse before vehicle applications came into service.

Apart from the mentioned lack of incentive towards higher power outputs, the turbocharger was initially regarded as inherently far from ideal for vehicle duties. It is a device with which, on vehicle engines, it is difficult to obtain sufficient air at low speeds without supplying too much at high speeds, with the consequent risk of excessive cylinder pressures. (In the case of marine and industrial applications, however this problem of matching is much less acute because the operating speed is constant for much of the time.) Consequently, when the vehicle industry began to show interest in turbocharging, the specialist manufacturers had to evolve a new technology to meet the major problems posed by these variations in the operating conditions, as well as those of coming down the size scale and keeping production costs to an acceptably low level.

One aspect of the problem was that the high inertia of the turbo-charger rotating assembly resulted in delayed response to the accelerator. This delay did not matter much on deceleration, but on acceleration heavy smoking tended to occur because the increased fuel delivery initiated by the governor was not immediately balanced by more air from the turbocharger. The makers therefore had to start thinking in terms not merely of scaling down their units proportionally to engine size but of designing much smaller ones that would have a low inertia but would achieve the required air throughput by a combination of improved efficiencies and very high rotational speeds. This approach gave mainly satisfactory results when applied in conjunction with modifications to the governor characteristics of the fuel-injection system.

Because of the matching difficulty, the automotive turbocharger makers had to evolve special methods of obtaining a torque curve compatible with vehicle requirements. Two such methods have come into common use. The first of these to be introduced was that of limiting the higher-speed output of the turbine by incorporating a 'waste-gate' in the exhaust manifold, upstream of the turbine. This waste-gate consists of a valve which on opening allows some of the exhaust gas to bypass the turbine; the valve is controlled by the turbocharger's boost pressure through the agency of an aneroid capsule.

The second method of flattening the torque curve was patented by Garrett AiResearch in the USA in 1963, and is to optimize the use made of the individual exhaust pulses at lower speeds. This objective is

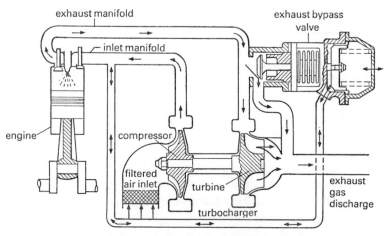

Schematic layout of a turbocharger with boost-controlled waste-gate, to limit the degree of supercharge, as developed for automotive diesel engines

achieved by dividing both the exhaust manifold and the turbine housing into two; one half of the manifold feeds the corresponding half of the housing. In the case of a four-cylinder engine, cylinders 1 and 4 deliver into one part of the manifold, and 2 and 3 into the other. On a 'six', though, the first three and the last three cylinders are grouped. In both cases, successive pulses alternate between one side and the other of the division, so interference is avoided. This arrangement too is used in conjunction with a boost-pressure-controlled aneroid capsule which operates on the governor of the fuel-injection pump. On purely technical grounds the pulse-energy scheme is preferable to the waste-gate, since the latter not only sacrifices some of the available exhaust energy when open but also tends to be more smoke-producing when the accelerator pedal is blipped. However, it is worth recording that recent work on very small automotive diesels (of down to 1 litre swept volume and running at up to 4500 rev./min.) has led to a revival of interest in the waste-gate as being simpler and cheaper to produce for engines of this size.

In earlier applications the turbochargers were designed to produce a relatively low maximum delivery pressure. As a result they could be added to existing naturally aspirated diesels without causing either mechanical or thermal overloading. However, when some engine builders began to look for further increases in specific outputs, demanding turbochargers of yet higher efficiency and greater pressure ratio, they were compelled to modify their designs in respect of such matters as crankcase and crankshaft stiffness, bearing capacity and the cooling of pistons and exhaust valves. In the case of some highly turbocharged vehicle engines, too, it has become necessary to cool the charge between the compressor and the inlet manifold, to obtain optimum power and fuel consumption and to minimize exhaust smoke in normal running. The charge coolers are produced by the heat-exchanger companies making radiators, oil coolers etc, and the air-to-water type is finding favour today at the expense of the air-to-air variety.

As was indicated earlier, minimal first cost of the turbocharger is an essential for automotive applications, since operators are seriously concerned with the purchase price of a vehicle as well as its overall running costs. Cheapening the turbocharger has had two aspects, the first of which is design and development to achieve the maximum efficiency from the smallest unit. The second is the evolution of more economical methods of producing the inherently expensive components – the compressor and turbine rotors and the latter's housing. Today, most vehicle turbocharger manufacturers make the two rotors very precisely by the 'lost wax' or investment casting process, respectively in aluminium alloy and a high-grade heat-resistant alloy, usually based on nickel. (Investment casting is covered in greater detail in Chapter 4, in the context of gas-turbine engines.) Turbine housing costs have been cut too where divided-flow is employed, by the elimination of the nozzle ring formerly incorporated, in favour of a volute-type nozzleless design which lends itself well to quantity production.

Valves and valve-seat inserts

VALVES

The mechanical purist has every right to apply the term 'barbarous' to the poppet-valve systems conventionally used to control the gas flow into and out of four-stroke reciprocating engines. Each valve is banged off its seat by a cam and snapped back on to it by a powerful spring, at frequencies of up to fifty times a second in quite ordinary car engines today, and around twice as high in racing power units. On top of that, the exhaust valve has to operate at very high temperatures, maybe even at dull-red heat when the engine is working really hard.

Yet, in spite of these physical disadvantages, the poppet valve – still basically the same as it was on the first vehicle engines back in the 1880s – is going as strongly as ever today. It has outlasted such apparently more civilized competitors as rotary and sleeve valves, though the latter admittedly had their years of glory in aero-engines before the advent of the gas turbine. That the poppet valve should have succeeded in spite of its manifest drawbacks is due partly to manufacturing economics and partly to persistent development through the years. Most of the progress has been metallurgical, for which the specialist manufacturers must share the credit with the metal laboratories, but there have been several important advances too in production engineering. Let us look first at the metals.

As mentioned in the first paragraph, the exhaust valve has much the harder life of the two. For many years, operating conditions were such that satisfactory results were obtained from a silicon-chromium steel generally known as Silchrome 1 and covered by British Standard Specification En52. Since the introduction of this alloy steel over forty-five years ago there have been many advances in engine design; compression ratios (and consequently working temperatures) have risen substantially through increases in fuel octane ratings and better combustion characteristics. The result of these improvements was that Silchrome 1 exhaust valves, in the more highly rated gasoline engines, began to show an increasing tendency to premature burn-out. Although the material was therefore superseded for those duties, it is still used today for the inlet valves of both gasoline and diesel engines, and for the exhaust valves in some naturally aspirated diesels.

Its main successor, which was developed in the late thirties, was XB – a chromium-nickel-silicon steel to BSS En59 – having a greater resistance to high-temperature corrosion and scaling. Steel of this type was quickly adopted by engine manufacturers and was long the accepted exhaust-valve material; even nowadays it gives satisfactory performance in engines of relatively low compression ratio. A minor breakthrough had been made, though, a few years earlier when Kayser Ellison – a well-known British steel firm of that era – brought out their KE965, the first 'austenitic' high-alloy steel, containing a substantial percentage of nickel plus chromium and tungsten. This proved more than adequate for the exhaust valves of high-performance engines, in both vehicles and aircraft, until several years after the 1939–45 war. However, it was of a higher grade than was necessary for more ordinary engines – hence the arrival of XB steel.

It is appropriate to interject here that, to give extra protection in engines having particularly severe operating conditions, exhaust valves have been available since the 1930s with a hard-facing on the seating area of the head and/or the tip of the stem. The material used for this hard-facing is Stellite, a cobalt–chromium–tungsten alloy. Another wear-reducing means adopted by the valve makers for some of their customers was to plate a thin 'flash' of chromium on the stems.

Around 1960 the continuing upward trend in compression ratios and power outputs again brought the exhaust valve into the familiar 'Achilles heel' situation. This was met by the metallurgists with the introduction of improved austenitic steels having a high content of chromium and manganese. The designation 21–4N was given to the first of these steels which had appreciably better hot strength and corrosion resistance than KE965. 21–4N was subsequently supplemented by 21–12N, a still higher-grade steel of the same type.

Even these materials soon became inadequate for particularly highly rated engines such as 'tuned' competition car units and turbocharged diesels. In such engines, which are often required to sustain high loadings for long periods, it is essential that the exhaust valve should not suffer from 'creep' – that is, permanent elongation under prolonged stress. To meet these especially arduous conditions, a nickel–chromium–cobalt alloy known as Nimonic 80 was adopted. It is one of a range of nickel alloys which, like Stellite, are not steels since they are not based on iron. As an indication of its excellent properties, Nimonic 80 is still used for the actual turbine blades in some gas-turbine engines. Reference to the use of hollow valves containing sodium, as a means of reducing head temperatures, will be found in Chapter 2, in the section on the cooling of aircraft engines.

Since the head of an exhaust valve is subjected to the highest thermal and mechanical loadings, and exotic materials are expensive, there is often a good case for using an austenitic steel for the head only, and having a stem of a cheaper martensitic steel. Apart from its price advantage, the latter can be selected to provide better thermal conductivity and bearing properties, and a lower rate of thermal expansion, with benefit to cooling, wear and clearance variation in the guide. The bimetallic valve has in fact become quite widely accepted during recent years and, of course, it stands or falls by the integrity of the joint between the two components. This brings us to one of the production engineering advances mentioned earlier: in the mid 1960s Farnborough Engineering Co., one of Britain's leading valve makers, were the first in the world to adopt the friction-welding technique for uniting head and stem. In friction welding one component is rapidly revolved while in contact with the other (thus generating considerable heat) and is then suddenly stopped while heavy pressure is applied to complete the weld. The technique has been proved to give a metallurgically and mechanically better joint than any other welding process – and the equipment costs are quite modest!

At about the same time Farnborough pioneered another production development, this time for increasing their output of valves with Stellite-faced seats. They evolved an automatic facing machine which handles six valves simultaneously, in sequence as the work-table revolves. Since each completed valve is replaced immediately by an untreated one, the process is continuous. The valves are loaded after their heads have been pre-machined to take the Stellite. On the machine each valve is rotated and its head is preheated before receiving its coating of molten Stellite; finally the head is reheated to alleviate any residual stresses and eliminate harmful oxides.

VALVE-SEAT INSERTS

During recent years there has been an increasing trend towards the use of aluminium instead of cast iron for the cylinder heads of gasoline engines for cars. Since further reference to this situation is made in the section on castings, forgings, etc., in Chapter 5 it is sufficient here to say that the main advantages of aluminium are its lower weight and higher thermal conductivity. However, its strength under compression is much lower than that of cast iron, so an aluminium cylinder head has to have valve-seat inserts of stronger metal (normally a cast iron) to avoid unduly rapid sinkage of the seats under the hammering action of the valves.

Being made in their thousands, these seats are of course very much in the province of the specialist supplier. A minor disadvantage of aluminium as a head material is that it has a considerably higher coefficient of expansion than cast iron. The initial practice regarding inserts, to minimize the risk of any loosening in service, was therefore to manufacture them of austenitic cast iron – a material that is more expansive on heating than ordinary cast iron. However, experience soon showed that seats of the latter metal gave no trouble if they were pressed into the head with a carefully chosen interference fit.

Then, in the late 1950s, advances in powder-metallurgy techniques caused Brico – one of the Associated Engineering companies and a well-known name in the manufacture of pistons and rings – to investigate the possibilities of sintered valve-seat inserts, made from powdered metal. These were found to be an excellent proposition, not merely in operational terms but also in respect of production costs: since they come from the moulds very close to their final dimensions, they need very little machining – to give the correct outside diameter and seat width only. During the succeeding years, Brico have evolved a range of iron alloys for these sintered seats, it being easy to vary the composition of the powder to give the optimum characteristics for the particular application. For example, exhaust valve seats usually contain a larger proportion of copper than do inlet seats, because it provides a denser and harder insert with better thermal conductivity. As an indication of Brico's abilities in the manufacture of sintered inserts, the Coventry firm's products are fitted to most British engines with aluminium heads, and to several built in other European countries.

It is worthy of interjection here that Brico's research in this field established the strong influence on seat life of the initial coaxiality between the valve head and the seating face. There are of course several 'areas of tolerance' here – namely the insert itself, its counterbore in the head, the bore for the valve guide and the guide itself. Brico maintain that the only way to ensure close coaxiality is to use the installed valve guide's bore as the datum for the positioning of the seating face, and many engine manufacturers now follow this recommendation.

There is a worldwide tendency to reduce the lead content of gasoline, for environmental reasons; in fact, completely lead-free fuels are now marketed in the USA and certain other countries. One of the side-effects of this lead reduction has been a significant rise in the incidence of premature sinkage of integral valve-seating faces in iron cylinder heads. At the earlier levels of lead compounds in the fuel, lead deposits would form on the seating faces and would serve both as a lubricant and as a protection against erosion.

Aluminium-head engines with sintered-iron seats seemed to be virtually unaffected by the sinkage problem. In 1971 Brico therefore carried out an experimental programme, in collaboration with Chrysler and Associated Octel (the main producers of lead-base anti-knock agents), to compare the performance of integral seats and sintered inserts in the cast-iron head of an engine run on lead-free gasoline. As was expected, the integral seats suffered rapid sinkage, whereas the inserts gave a very satisfactory life. The enterprise was not crowned with commercial success, though, because in the USA – the first country to impose restrictions on lead content – the alternative measure of induction-hardening integral seats had proved successful a little earlier, and was cheaper than fitting inserts.

The foregoing remarks on valve seats apply of course to vehicle gasoline engines only. Because of the tougher thermal conditions in most diesel power units in this field, and operators' insistence on a long life between overhauls, it has for many years been the practice to fit these engines with inserted seats of special cast iron. The metallurgical boffins have evolved a range of alloy irons for this purpose; the alloys usually contain silicon, manganese and chromium – plus either nickel or molybdenum – and the actual choice depends on the valve material, engine details and operating conditions. Endurance testing on the dynamometer is still the best way to establish the most compatible seating alloy.

As would be expected, though, sintered inserts have come under close scrutiny for diesels owing to their success in gasoline engines. A lot of development and test work has been done recently on suitable alloys by Brico and others, and the results have been very encouraging from the durability aspect. It therefore seems likely that sintered seats will come into general use within the next few years, at least in the smaller, higher-speed automotive diesels.

Camshaft and other drives

The early atmospheric gas engines, being derived from steam power units, had the inlet and exhaust phases controlled by slide valves. However, once the four-stroke cycle became established and speeds began to rise, the slide valve was too cumbersome and was soon replaced by the mushroom-shape poppet valve. For quite a number of years it was standard practice for the inlet valve to have automatic operation, opening when atmospheric pressure exceeded cylinder pressure as the piston descended on the induction stroke and closing when the converse applied at the beginning of the compression stroke. The exhaust valve, though, had to be mechanically operated since it had to open against a positive cylinder pressure, and the obvious method of operation consisted of a half-speed camshaft and a linkage system to the valve.

Spur gearing with straight teeth was used originally to drive the camshaft from the crankshaft. These gears were inherently noisy, and the increasing demand for refinement soon led to the adoption of helical gears which sometimes were even hand-scraped to ensure correct meshing. A well designed, manufactured and installed drive of this type was efficient and quiet when new, but was expensive to produce and tended to get considerably noisier as wear occurred.

Consequently it was not long before engine manufacturers began to investigate the possibility of replacing the gears by a specialist-made chain and sprockets. The first recorded example of such a drive was on the Knight engine introduced in 1904. Other makers took up the principle, and when the magneto came into vogue for ignition purposes the drive was triangulated to operate it also. The so-called 'silent' or inverted-tooth chain, which runs on gear-type sprockets, was the only design used until about the middle of the second decade of this century. Then a few manufacturers of small engines started to employ the cheaper, more compact roller chain which had been invented by Hans Renold in 1880 and had quickly been adopted for bicycle transmissions.

Shortly after the end of the First World War, when short-pitch duplex and even triplex roller chains had been developed by Renold and other specialist companies, this type of chain started to make rapid inroads into the camshaft-drive market in Britain and Europe. The reasons were the already mentioned lower cost and reduced width for a given transmitting capacity, together with manufacturing improvements that brought the noise level almost down to that of the inverted-tooth chain. A further advantage of the roller chain was that it lent itself more readily to the incorporation of an automatic tensioner, thus extending the life of the drive. Modern adjusters of this type, as developed by the chain specialists, are operated by engine oil pressure and occupy very little space, as well as costing remarkably little. Although the roller chain became almost standard on this side of the Atlantic during the 1920s and 1930s, the inverted-tooth type remained the choice of US engine builders and has continued so to the present day.

Before we leave chain drives it should be mentioned that the roller variety has a long record of service on motorcycles, for the primary transmission from engine to gearbox, as well as for driving magnetos and/or dynamos. Since the First World War, various motorcycle engines have been built with chain-driven overhead camshafts, but gears have always been preferred for side-valve and pushrod o.h.v. units because of their extreme compactness which is more important in the two-wheeler context than for other vehicles.

A new contender for camshaft and auxiliary drives – the toothed

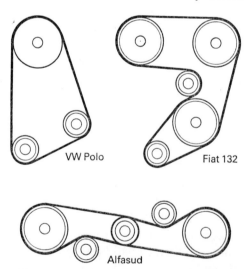

Three of the many possible configurations of toothed-belt camshaft drives for car engines; the VW Polo and Fiat 132 are in-line units while the Alfasud is a flat-four

belt – appeared on the automotive scene during the late 1950s and has now become an acceptable alternative to the chain in a number of applications, particularly where two overhead camshafts are used to actuate the valves. This type of belt was invented in the USA in 1940 by a Uniroyal employee and its original application was to 'timing' drives on complex industrial machinery. It consists essentially of a virtually inextensible strength member, comprising helically wound cords (usually of steel or glass fibre), embedded in synthetic rubber on which the teeth are formed. The pulleys are of course similarly toothed to mesh with the belt.

In comparison with the chain drive for camshafts, the toothed belt system is rather cheaper and requires no lubrication, so its enclosure can be of simpler nature. Also, since its pitch length does not extend with wear, it does not have to have any tensioning arrangement, though this may be provided to facilitate installation. However, the belt is generally less durable than the chain and has to be wider for a given power-transmitting capability, thereby increasing the length of the engine. Apart from its application to camshaft drives, the toothed belt has come into widespread service for other drives such as those to petrol-injection metering units and to hydraulic pumps for steering power-assistance, where slip (as may occur with vee-belts) cannot be tolerated.

Ignition equipment

Gasoline engines, in contrast with the diesel or compression-ignition variety, have the fundamental requirement that the combustible mixture in the cylinder has to be ignited deliberately – it does not catch fire spontaneously. Many students of engine history seem to believe that the early gas engines had only hot-tube ignition systems, featuring a metal or ceramic tube in the cylinder head; the closed outer end of the tube was heated by an external flame and the open inner end got hot enough to ignite the charge. However, Lenoir's 1860 gas engine had electric ignition by means of a proprietary high-tension trembler coil and a sparking plug, the latter bearing a marked resemblance to those used today.

Hot-tube ignition *was* used on numerous early vehicle engines, including those built by Daimler and Panhard, but a changeover to electric ignition was beginning by 1890 and had got well under way by the very early 1900s. The first vehicle systems were of the low-tension type consisting of a battery and a sparking plug involving a contact-breaker arrangement situated within the combustion space. Although this scheme worked reasonably well, it had the disadvantage that moving parts had to operate in a hot and corrosive environment.

The next stage of development can be attributed to the de Dion Bouton car firm whose engineers decided at square one to adopt high-tension ignition. They started with a trembler coil layout but then, in the search for higher engine speeds and power outputs, switched to a single-break system involving an LT/HT coil actuated by a mechanically operated contact-breaker – the precursor of the type of equipment that subsequently became generally used throughout the world. One reason for the success of this ignition scheme was that, by then, coil manufacturers were able to build units robust enough to withstand the vibration and exposure common to vehicles at the turn of the century.

In spite of this pioneering work by de Dion Bouton, the trembler coil was in widespread use for a number of years on vehicle engines. Its eventual slow eclipse was in fact brought about by the magneto which had been under development, primarily by Simms in Britain and Bosch in Germany, since about 1895. The advantages of the magneto over other systems were that it was self-contained (it needs no battery, of course) as well as compact and reasonably reliable. Early magnetos were of the low-tension variety, used in conjunction with the contact-breaker sparking plug already mentioned. Then Simms hit upon the idea of feeding a separate high-tension coil from the LT magneto, but the Bosch company went farther by combining the LT and HT windings in one armature to produce the first practicable high-tension magneto. This actually went into production as early as 1901 but did not gain much popularity for several years.

By shortly before the outbreak of the First World War, Bosch had become much the largest manufacturer of HT magnetos for automotive purposes, for Britain and the USA as well as Europe. At that time, though, a brilliant American vehicle engineer, Charles F. Kettering of General Motors (later to become Director of Research), decided that the battery/coil system warranted resuscitation, in the single-spark form, of course. His reasons were threefold – easier engine starting, lower cost and the improvements that had been made to batteries. The easier starting is due to the fact that the coil generates its highest voltage at low engine speeds, whereas the HT output of the magneto rises with speed.

As a result of Kettering's endeavours, the early post-war generation of American cars had coil ignition, but not until the late 1920s did this system begin to find real favour, on its own, in other countries. However, dual ignition – comprising a coil for easy starting and a magneto for normal running – was taken up relatively early by a number of makers of 'quality' cars (Rolls-Royce and Panhard, for example) and continued to be used in some instances for quite a long period after the war. Some car manufacturers produced their own ignition equipment during this formative period, but gradually changed to proprietary units as the specialist firms became more competent. By the early 1930s, coil-ignition sets made by these companies – Bosch, Lucas, Delco and others – had become standard vehicle equipment.

It was appreciated before the end of the last century that the timing of the spark in relation to the piston position had to be variable. The timing that enabled the engine to be started without a vicious kick-back through over-advance was too retarded once a reasonable running speed had been attained. Control of the contact-breaker timing by the driver through a mechanical linkage was the first solution and one that continued for many years. Then in the early 1930s, the electrical manu-facturers began to incorporate centrifugal automatic control, by means of bobweights in the contact-breaker unit. This arrangement ensured adequate retard for starting and slow-running, and advanced the timing progressively over a predetermined range as engine speed increased. However, it took no account of the need to retard the ignition on hard acceleration, to avoid detonation.

The final stage – reached shortly before the Second World War – was therefore the addition of a vacuum control whereby the ignition timing was automatically retarded when the throttle was opened, through the consequently reduced depression in the inlet manifold. Here we have a typical example of the frequent need of close collaboration between the manufacturers of the engine and of its ancillaries, since it is not easy to obtain the correct blending of the two automatic control systems in relation to the optimum advance requirements of individual power units.

It should perhaps be mentioned here that for motorcycle engines the pattern of change differed from that in other vehicle categories. Because most riders were enthusiasts, and took an interest in the finer points of controlling their machines, magneto ignition with manual advance/retard was widely preferred to the coil system until the 1950s, when the alternator began to replace the not-too-reliable dynamo for battery-charging purposes, and batteries themselves became more robust. Coil ignition is now in general use on motorcycle engines but automatic control of the timing is of the straightforward centrigugal type only.

In spite of its popularity, the orthodox coil-ignition system of today cannot be regarded as the ultimate. Its limitations were first underlined in racing when, during the 1960s, the new generation of V8 and even V12 engines required 40000 or more sparks per minute. Such a rate was bad enough for the coil manufacturers but could be met by duplication; the real weakness lay in the mechanically operated contact-breaker which, because of its inertia, suffers at high speed from 'bounce', thus reducing the closed period of the contacts and hence the energy stored; this phenomenon sets a practical limit of about 24000 sparks a minute. In addition, the contact-breaker gap (which itself affects the timing) varies with wear of the cam follower heel and erosion of the points themselves through burning.

The ignition specialists were able to tackle these problems by use of the up-and-coming technology of electronics. There were two stages to the development of electronic ignition systems. In the first of these the original contact-breaker was retained, but only as a base-line switch transmitting the very small current required to 'trigger' a transistor that handled the actual primary current for the coil. This arrangement was not expensive and certainly reduced erosion of the points, since the current was too small to cause significant burning, but the other disadvantages of the contact-breaker remained. The full solution lay in the second stage – the more costly electronic systems from which the contact-breaker was eliminated.

Various systems of this type are now on the market. Their details are outside the scope of this book but the principle is worth outlining. Again a transistor handles the coil's primary current but the triggering pulse for it is generated at the correct frequency and timing by an opto-electronic or electromagnetic device; one such device is a discontinuous rotor revolving between two magnetic pole-pieces. These more advanced electronic ignition systems are now in general use for racing and other competition purposes and are beginning to be adopted for ordinary road cars and motorcycles, while proprietary conversion sets are available for fitting by the enthusiastic owner.

Although electronic equipment may effect little improvement in the ultimate performance of engines with modest maximum sparking rates, it does obviate the routine adjustment and replacement of the contact-breaker set while maintaining constant spark timing. Also, its more consistent performance, not merely at high speeds, has been found to reduce the unburnt-hydrocarbon content of exhaust gases because of the fewer misfires that occur. It follows that the emphasis on reducing pollution could eventually force all engine manufacturers to adopt superior ignition systems.

Another relatively recent variation on the ignition theme is what is called the capacity-discharge system, in which the primary side of the coil is fed from a capacitor (condenser) rather than directly from the battery. The resulting considerable shortening of the 'rise time' of the secondary voltage ensures sparking under plug-fouling conditions too severe for a conventional coil/contact-breaker set-up. However, this advantage is offset by the spark's much-reduced duration which impairs the ability to ignite weak mixtures. Capacity-discharge systems are consequently unsuitable where exhaust emissions are critical, unless the spark duration is boosted by the incorporation of additional electronic equipment which raises the cost substantially.

The history of that small but vital component, the sparking plug, has of course gone hand in hand with that of HT ignition systems. Basic principles have remained virtually unchanged through the years (even the original de Dion Bouton plug had the 14 mm thread that is the most common today) but there have been many improvements as the plug manufacturers have had to meet ever more stringent requirements. The first examples had central insulators of porcelain which had poor resistance to thermal and mechanical shock. As power outputs rose, so did the stressing on the insulators; physical failure therefore became common, particularly in competition work, and many an engine was wrecked as a result.

Such failures led Kenelm Lee Guinness, a famous racing driver of the era before the First World War, to experiment in 1911 with mica as a replacement for porcelain, with such success that he founded his own KLG company to produce mica-insulated plugs. The material was duly adopted by numerous other manufacturers and proved satisfactory until the introduction in the late 1920s of tetraethyl lead as an anti-knock agent in gasoline. Mica was found to be attacked by the products of combustion of lead-containing fuels, so a replacement had to be found. Various well-known firms in Britain, Germany and the USA investigated numerous alternatives, mostly of a ceramic nature, and the most satisfactory – in terms of thermal and mechanical strength and ease of manufacture – proved to be alumina, or aluminium oxide, which is still the best today. A more detailed reference to this material will be found in the next chapter.

As engine design advanced, the plug companies had also to evolve improved materials for the central and earth electrodes, to enable these to resist erosion by the products of combustion in the cylinders. Most earlier plugs could be dismantled for cleaning off deposits and furbishing the electrodes. However, there was always the risk of subsequent mal-sealing through damage or the inclusion of dirt; fortunately the combination of more durable electrode materials and better control of the lubricating oil (the burning of which was responsible for much of the deposits) enabled the manufacturers to go over to 'one-piece' construction which has been used for virtually all automotive sparking plugs since the Second World War.

A problem that has always confronted the plug makers has been to design a component that, in a particular engine, would not get its 'business end' either hot enough to cause pre-ignition (and the risk of holed pistons) or cool enough to result in fouling by combustion products which should be burnt off lest they cause misfiring. The difficulty of striking the correct balance between heat input and dissipation has become greater of recent years because of the widening range of operating conditions encountered by many vehicle engines. Nevertheless, by careful attention to insulator nose design and to heat paths, the plug manufacturers have usually managed to satisfy most of their users most of the time.

In this connection, developments in combustion chambers and valve gear have tended to increase the physical difficulty of finding room to install the sparking plugs in the optimum position. To ease this situation, scaled-down (10 mm) plugs were introduced around 1950 and they worked well (and still do) in competition car engines and some motorcycle units. However, they proved less satisfactory in normal road-going car engines because the smaller the bottom half of the plug the narrower the heat range – and consequently the greater the risk of either preignition or fouling.

One ingenious solution to the dilemma is the taper-seated plug pioneered in 1955 by Champion, in conjunction with Ford, in the USA. It was first adopted on this side of the Atlantic in 1968 by Vauxhall Motors, for their then-new overhead-camshaft engine, and has since been taken up by several other manufacturers. In this design the quite wide square shoulder and sealing washer above the thread are replaced by a conical shoulder which pulls down directly into a matching recess in the cylinder head. Because of the narrowness of the shoulder, the thread and lower half of the body can be large enough to give an adequately wide heat range, while the hexagon and upper part of the insulator can be relatively small and therefore easier to accommodate – the best of both worlds, in fact.

Less successful so far have been the efforts to establish the surface-discharge sparking plug. As the name suggests, the spark of such a plug travels from one electrode to the other along the surface of a solid dielectric (filling the space between them) and not across an air-gap. According to theory, surface-discharge plugs should overcome the problem of misfiring through fouling, since deposits provide a conductive path for the electricity. In practice, though, conventional vehicle ignition systems do not generate either a high enough voltage or a short enough voltage 'rise-time' to enable such plugs to work properly in all conditions. For these combined reasons the use of surface-discharge plugs is virtually confined to high-output American outboard motors with capacity-discharge ignition which does create the right conditions. In Chapter 4, though, reference is made to a very similar device used for initiating the continuous combustion in gas turbines.

Perfection on the ignition side has yet to be reached, not least because of the way in which operating conditions are liable to variation; each change can produce new problems. The latest crop of difficulties arise from the double emphasis on low exhaust-gas toxicity *and* low fuel consumption. This combination is forcing engine researchers to experiment with air/fuel mixtures as weak as 18 or 20 to 1 (or even beyond), and such mixtures are very difficult to ignite and keep burning. Consequently, still more advanced ignition equipment is under development in the laboratories of the specialists. Among the possibilities being studied are multiple-sparking arrangements and high-energy systems which will necessitate plug electrodes of precious metal to combat erosion.

Starters and generators

Until 1911, the only standard electrical equipment on motor vehicles was the engine's ignition system. As already discussed, the high-tension magneto was at that time the standard ignition means, since it was self-contained and did not require a battery. Engines had to be started, as from the earliest days, by turning them over with a handle or pedal – an irksome, uncertain and sometimes quite injurious method! In that year, though, Charles F. Kettering (mentioned previously in connection with coil ignition), of General Motors, invented the electric starter and fitted one experimentally to a Cadillac. By so doing, he probably did more towards popularizing the motor car than anyone else had done hitherto.

Happily, Kettering's invention proved an immediate success, and the GM organization must have worked fast to get production versions of his starter (made by Dayton Engineering Laboratories Co., later to become part of GM as Delco) on to the 1912 Cadillac models. The electric starter of course necessitated a battery – fortunately already there on Cadillacs for coil ignition – and this in turn had to have a generator to keep it charged, so here we have the US beginning of the vehicle electrical system as we understand it today. In the UK, on the other hand, the introduction of the generator was associated with the coming of electric lighting, for which 'add-on' dynamos were available from Lucas as early as 1910.

Kettering's electric-starter scheme was simple and logical: he designed a combined motor and generator which, by means of ingenious switch-gear, operated on 24 volts to start the engine and then, when this was running, charged the battery at 6 volts. The dual-purpose electrical machine, commonly called a dynamotor, was subsequently taken up by a number of other companies, in various forms, and lasted into the 1930s on car engines; in the 1950s, though, it had a relatively short revival on motorcycle-type engines. It had the basic disadvantage of being less efficient in either function than a single-purpose unit – hence the increasing preference of vehicle and electrical manufacturers for separate starter motors and generators.

With this arrangement, of course, the starter did not need to be designed for continuous running – merely to provide a high output for a limited period. However, it therefore had to incorporate some means of disconnecting it from the engine once it had performed its job. Rather surprisingly, one of the first methods evolved to disconnect automatically – the Bendix inertia drive, patented in 1912 and first employed in about 1914 – is still in use today, although basically it is a mechanically barbarous arrangement. The drive comprises a ring-gear on the engine flywheel and a quickthread-mounted pinion on the starter-motor spindle. When the motor is energized, the pinion winds itself along the quickthread into engagement with the ring-gear; once the engine starts, though, the acceleration of the pinion causes it to wind itself back out of mesh.

The Bendix drive is still fitted to many British quantity-produced cars and works satisfactorily on them in temperate climates such as ours.

However, it can suffer from premature ejection of the pinion on gasoline engines having high compression ratios and/or small flywheels, or where automatic transmission is fitted or low starting temperatures are encountered. The answer in these cases lay in the starter with pre-engaged drive, first developed in the USA shortly before the Second World War. Starters of this type are now fitted by most foreign car manufacturers and by British ones on some models, particularly for export; they are invariably used on automotive diesel engines. In the pre-engaged system, the pinion is brought into mesh with the ring-gear before the motor is energized. A small over-run clutch between the pinion and the motor spindle prevents the engine from driving the motor (at relatively high speed) after starting, before the pinion has come back out of engagement. The scheme has the secondary benefit of reducing starter noise.

Because batteries, coil ignition systems and starter motors require direct current, it was entirely reasonable that the first generators for vehicles should have been of the type known as dynamos, having a DC output. In their original form, dynamos had merely two carbon brushes to collect the current from the commutator at the end of the rotating armature, and a fixed third one to supply the 'field' current to the stationary electro-magnets. They had no form of output control, which meant that the battery tended to be undercharged in winter, when the starter and lights (if fitted) were used more, and to be overcharged in summer.

During the 1920s the electrical-equipment manufacturers produced two relatively simple methods of improving this situation. One was two-position, half-charge/full-charge switching; in the half-charge position of the switch, a resistance was brought into the circuit to reduce the charging current fed to the battery. The other scheme was called third-brush control: the third brush on the commutator was made to be manually adjustable for position, to provide a limited variation in the dynamo output. It was therefore possible, though with a bit of trouble, to compensate for undercharging or overcharging conditions of operation.

Clearly, neither of these schemes was really satisfactory since they involved the human element. What was needed was fully automatic control of both the voltage and the current delivered by the generator, so the electrical manufacturers got down to this requirement during the 1930s. The resulting systems, developed by such well-known companies as Lucas, Bosch and Delco among others, were known as AVC (automatic voltage control) units but they also incorporated the necessary current regulation. In effect they were derived from the electromagnetic cut-out which had been incorporated in dynamo installations from the earliest days; its purpose was to disconnect the battery from the dynamo when this was at rest or being driven too slowly to provide charging current, thus preventing the battery from discharging itself through the dynamo. On vehicles that still have DC generators, the control units fitted today are still very similar to the original designs.

As can be inferred from that last comment, the big majority of automotive engines that have to produce electricity for 'home consumption' now have generators with an AC output, these being known as alternators. The basic reason for this preference is that the dynamo, with its rotating armature and commutator, and its brush-gear to collect the current, is not so robust either electrically or mechanically as the alternator which has a rotating magnetic field and stationary windings in which the current is generated. An analogous situation will be found later in the book, in connection with aircraft magnetos.

In fact the first production alternators (by Lucas) were fitted to motorcycle engines in the early 1950s, at a time when coil ignition was beginning to replace the previously almost universal magneto on two-wheelers. These alternators were very simple in concept, yet they worked well enough in practice. The rotor was merely a multi-pole permanent magnet with no excitation windings, and there was no voltage or current regulation – merely a switch giving low-charge or high-charge output. However, there was, ingeniously, an additional switch position which channelled the entire alternator output to the ignition coil, bypassing the battery, so that the engine could be kick-started despite a flat battery. The AC generated had of course to be converted to DC for ignition and battery-charging purposes, and this was achieved by the use of rectifiers, which were initially of the selenium type and later of the more efficient silicon type.

From those simple and quite inexpensive motorcycle alternators have sprung families of more advanced automotive units, with such refinements as rotor field coils and fan cooling, and inbuilt rectification of the generated current. Outputs of several hundred watts are available making possible the large quantities of highly consumptive electrical equipment – such as heated rear windows and multi-lamp installations, for instance – now fitted to many vehicles. In this context, not the least advantage of the alternator is that, because of its greater robustness, it can be run at a higher speed than a dynamo. Consequently it can produce a useful charge even at engine idling speeds; with a D C generator at these speeds, any electrical equipment in use represents a direct drain on the battery.

A lot of work was done by the specialist manufacturers during the 1950s and 1960s on control systems for alternators. In that respect, these machines are simpler than dynamos in that they are self-limiting on current output and so need only voltage regulation. Thanks to transistors, and other miniaturized components, it has now become possible to house the voltage regulation system within the casing of the alternator.

In closing this section, a brief mention should be made of the methods used to drive generators of both kinds. On most motorcycle engines, dynamos were driven by gears or chain from the timing train, but alternators have always had the rotor mounted directly on the end of the crankshaft. The standard practice for cars and larger vehicles, though, has long been to use vee-belts and pulleys, the power again being taken from the crankshaft nose. During the years the vee-belt manufacturers have made substantial progress in a relatively unassuming way. By the adoption of advanced materials and manufacturing methods, they have progressively reduced the belt section necessary for a particular duty. The smaller section has in turn enabled smaller pulleys to be used without risk of overheating of the belt through excessive flexing, so drives are nowadays considerably more compact, as well as more durable, than they were.

Radiators, thermostats and fans

Since the majority of reciprocating engines (excluding motorcycle units) and Wankel rotary engines are water-cooled, some means has to be provided of dissipating the waste heat of combustion. The radiator – really a misnomer since most of its heat dissipation is by convection and conduction – is therefore an essential part of an engine installation, although it cannot be regarded strictly as an engine 'accessory'. The relationship between occupied space, weight and cost for automotive radiator systems differs from that for the cooling arrangements of other types of engine. There is some basic similarity, though, between automotive and aircraft requirements: installation space is almost invariably restricted in motor vehicles, and undue weight is a disadvantage, while low frontal area and minimal weight are essentials for aircraft.

From the earliest days of the vehicle engine, copper and brass were the orthodox materials for radiator construction, brass of course being an alloy of copper and zinc – added to increase the strength. The reason for this choice was the high thermal conductivity of copper, a characteristic that enabled the heat in the water within the radiator to be transmitted at a high rate into contact with the air outside. Some of the first radiators were merely lengths of relatively large-diameter copper pipe, convoluted to give a reasonably big surface area in relation to the occupied space. Such items could of course be readily produced by the vehicle manufacturer, but quite soon a specialist branch of the industry began to develop, as engine powers rose and with them the amount of heat to be dissipated.

To avoid having an excessively large radiator, the surface area had to be increased, and this was first done by soldering-on fins of either annular or helical form. It was not long before the single convoluted tube gave way to the series of straight tubes soldered into brass top and bottom tanks. Some radiators of this type had separate finning on each tube, while in others the fins comprised thin brass or copper sheets pierced to take a number or all of the tubes. At much the same time as finned-tube designs were being developed, the so-called honeycomb radiator came on the scene. It was built-up from a series of short longitudinal tubes having hexagonally formed ends with a circular section between them; the ends of adjacent tubes were soldered together, and the water flowed through the spaces between the circular portions. Radiators of this kind had a good appearance and were used on cars for many years, but were costly to produce and liable to leak with any maltreatment.

Then in the early 1930s came the beginning of the trend towards the decorative frontal grille, which enabled the radiator manufacturers to go over to cheaper and more robust forms of construction, and to improve cooling efficiency. One layout, which is still widely used today, was in fact derived from the earlier finned-tube design; the tubes are flattened, however, to give them a more favourable ratio of heat-transfer area to water weight than has a circular-section tube. In most instances the finning was of the continuous plate type, but some manufacturers began to make radiators with fins of corrugated shape between adjacent rows of tubes. In the 1940s an alternative design, which provided a very good ratio of area to water weight, was the 'film-tube' radiator. In this, the water passages and air passages are formed integrally in the radiator core which comprises a series of ingeniously shaped serpentine strips sweated and interlocked together to give a cellular construction. However, this layout has the same disadvantages as had the earlier honeycomb type.

Until recently it was standard practice, regardless of the detail design, to have a vertical water flow through the radiator. The trend towards wider and lower bonnet lines on cars, however, has caused some manufacturers to adopt 'crossflow' radiators in which the flow is horizontal. Such a layout means that no assistance to the flow is given by thermosyphon effect, but this is no longer of any significance because of the general adoption of water pumps to ensure an adequate rate of circulation.

For many years radiator manufacturers have been interested in finding alternatives for the traditional copper/brass construction, primarily to reduce the cost/heat-dissipation ratio. In fin-and-tube radiators particularly, the fins (and even the complete assembly) were sometimes made of steel, and various other materials were investigated with varying degrees of success. There was once a Japanese radiator with paper fins; it was a special sort of paper, of course, impregnated with copper dust to improve the conductivity!

However, aluminium eventually came to be regarded as technically the most promising alternative, but it was not an economic possibility until the late 1960s. By then, the price of copper had risen significantly more than that of aluminium, so the 'crossover point' was reached where the higher cost of aluminium construction had become at least offset by the lower material price. The thermal conductivity per unit volume of aluminium is inferior to that of copper, but it was felt that careful design of the core could result in a radiator that was no bulkier than one of conventional materials – for the same cooling performance – while being significantly lighter as well as cheaper. That this assessment was accurate is witnessed by the fact that already several current-production cars are fitted with aluminium radiators, and there is every indication that their numbers will increase substantially in the future.

Three methods of constructing aluminium radiators have been evolved by the specialist companies concerned. One is brazing, using a fluxless technique to obviate the corrosion risk that arises where a flux is used,s unles an expensive process is used to remove the latter. (It is worth interpolating here that one or two aluminium radiators were in production in the 1950s, but they were built up by fluxed brazing and the cost of removing the flux soon led to a loss of interest in further development.) The second method of construction, perfected by the French Sofica firm, is mechanical jointing. In this technique, the water-carrying tubes are expanded into fin plates, and into end plates comprising the bases of the top and bottom tanks which can be of metal or plastic material.

The third process is adhesive bonding, pioneered by Associated Engineering in Britain. As the term suggests, the various components of the radiator are stuck together by a synthetic resin which is cured by stoving. This technique gives a strong construction and appears to lend itself well to large-quantity automated production. Its theoretical disadvantage is that the layers of adhesive between the components have an insulating effect which reduces the rate of heat conduction. To overcome this difficulty, AE evolved a special design of core giving an unusually large area of contact between waterways and fins. It is regrettable that, for commercial reasons, AE recently had to shelve this interesting radiator which was to have been put into production by Covrad, one of their associated companies.

Because of the load and speed variations of a vehicle engine, and the frequent changes of air speed, the temperatures in an uncontrolled cooling system vary considerably. However, engines themselves operate most efficiently over a relatively narrow range of coolant temperatures. The importance of some form of temperature control began to be appreciated in the 1920s, and the first method of obtaining such control was the fitting of shutters to the radiator; the degree of opening of these shutters was controlled either manually (in conjuction with a temperature gauge) or automatically by means of a thermostat.

Then in the 1930s, when pump-assisted circulation of the water was gaining wide acceptance, thermostats began to be used to control the water flow from the engine to the radiator, and hence the cooling effect. The production of these thermostats was of course in the province of the specialist manufacturer. For many years the 'bellows' thermostat was the standard type, and it still has limited use today. Its working portion comprises a concertina-like capsule (or bellows) filled with a volatile liquid which evaporates on heating; the consequent vapour pressure causes the capsule to expand axially, the movement being employed to open a valve in the main coolant passage. Above a predetermined temperature, therefore, the water is allowed to circulate through the radiator, but below that temperature the circulation is within the engine only.

A post-war development has been the wax-element thermostat which is basically more robust than the bellows type and can give more precise temperature control. In the wax-element device use is made of the high coefficient of thermal expansion of microcrystalline paraffin waxes. A pellet of such wax is contained in a small capsule housing a pencil-shape rod with a protruding end. Expansion of the wax on heating tends to eject the rod from the capsule, by an amount proportional to the temperature rise, and the relative movement between the two components is harnessed to operate the water valve against the resistance of a spring. The wax-element thermostat first found favour in the USA and was soon taken up in other vehicle-manufacturing countries; because of its already-mentioned operational advantages, and the fact that the design lends itself readily to large-scale, low-cost production, it has become the dominant type in the vehicle field.

At quite an early stage in the history of the vehicle engine, once the matrix-type radiator had established itself, a fan was found to be usually necessary to assist the flow of cooling air. The obvious method of driving the fan was from the engine itself; belt drive from the crankshaft was normally employed, but sometimes the fan was mounted directly on the crankshaft. It later became almost standard practice to have the fan in tandem with a water pump housed on (or in) the engine block, the pair being driven by a vee-belt which also drove the generator once the latter had come into general service.

Many of the earlier fans were aluminium castings but when engines began to be mass-produced by numerous manufacturers, by the late 1920s, cost-consciousness reared its rather ugly head, so cheaper means of production had to be found. Pressed-steel fans therefore came into use on the majority of automotive power units and lasted well into the post-war era – even until today, in some cases. Many of these fans were decidedly 'cheap and nasty', not least for having a constant angle of twist right along the blades – a feature that hardly made for air-passing efficiency.

A change for the better started in the late 1950s, thanks to advances in plastics technology. Moulded one-piece fans, usually of polypropylene, began to find favour, particularly for cars and light commercial vehicles. By using injection moulding the plastics companies were able to produce blades of proper aerodynamic form, with the correct variation of pitch angle from root to tip. It became possible also to have many more blades (two or four had been the most common before), so not only could the diameter be reduced but the efficiency rose further and the noise level came down.

Of recent years, efficiency has been enhanced in some instances by ducting the fan to improve the airflow through it. However good a fan is, though, power is still consumed in driving it, and this power is wasted unless the engine is working hard enough for the fan to be essential to meet the heat-dissipation requirements. In many operating conditions the fan is superfluous for 90 per cent – or even more – of the total running time; a lot of fuel can therefore be squandered in driving it unnecessarily.

Numerous methods of countering this waste have been evolved by the cooling specialists and firms in the accessories field. The simplest is probably the variable-pitch fan in which, as the air speed rises, the pitch of the blades is automatically fined-off (by the aerodynamic forces or the corresponding torsional loads overcoming spring pressure) to reduce the air-passing ability and hence the power consumption. Another approach is to limit the fan speed by fitting a 'viscous coupling' between the driving hub and the fan itself; the two members of the coupling are separated by a silicone fluid which allows slip to occur progressively (because of the drag of the fan) once the hub speed exceeds a predetermined figure. Such couplings have achieved some popularity in Britain (where they have been made for some years by Holset Engineering) as well as in other countries.

Speed-limiting is only a palliative, however; the best technique is to disconnect the fan completely when it is not needed, the disconnection and recoupling being effected thermostatically from the coolant temperature. An interesting method of doing this was the magnetic-particle coupling introduced in Britain by Smiths Industries in 1960. Between the driving and driven members of this coupling was a space containing the magnetic particles. The driving member incorporated a solenoid, activation of which by an electric current caused the particles to orientate themselves so as to form a virtually solid bridge from one member to the other, so the two revolved together; when the current was switched off, the bridge collapsed. The switching operations were done by simple thermostatic control from the coolant.

Despite the ingenuity of this device, it has today been superseded by electric or hydraulic drive – the latter for the larger automotive diesel engines. Both schemes have the advantage of flexibility of installation, in that the power source can be remote from the actual fan. The electric unit has merely an on-off control whereas the more sophisticated hydraulic drives are speed-controlled from zero to maximum according to the temperature conditions. Consequently, hydraulic drive lends itself particularly well to the compact 'packaged' cooling systems that have been developed during the last few years for vehicles such as tanks in which space is at a premium.

2 Aircraft reciprocating power units

Introduction

Although the Wright brothers are believed to have built the first aircraft engine that actually flew – at Kitty Hawk, USA, on 17 December 1903 – there are records of earlier less fortunate power units for heavier-than-air machines. A 50 h.p. water-cooled radial was designed a year or two earlier, also in the USA, by C. M. Manly for Professor S. P. Langley, but Langley's aircraft crashed on launching and wrecked it. At much the same time, the Antoinette company in France, convinced that flying was 'on the doorstep', built a 24 h.p. V8 in readiness, although it was first used for motor-boat racing, not aviation.

The Wrights' engine, which reputedly developed a mere 15 h.p. at 850 rev./min. and weighed 240 lb, was a horizontal four-cylinder with an aluminium crankcase, automatic inlet valves, a surface carburettor and low-tension ignition. It was, in fact, the sort of unit that would have been expected from the state of knowledge at that time. Two years later Voisin built an aircraft for the Brazilian pioneer Santos Dumont and fitted it with a larger, 50 h.p. Antoinette engine. This machine is considered to have made the world's first fully controlled flight, towards the end of 1906, but it seems astonishing today that none of those involved apparently knew about Orville Wright's success until over a year later, when the brothers came to Europe. There Bollée built them a bigger, 50 h.p. engine which weighed no more than their original one and was capable of running at 1450 rev./min. It had a form of manifold injection for the petrol, high-tension ignition and auxiliary piston-controlled exhaust ports.

From 1908, progress in aero-engines was rapid, since flying had captured so many people's imagination. An entirely new type – the rotary with radially disposed cylinders – was built by Gnôme in France during 1909, and immediately caught on because of its light weight (around $3\frac{1}{2}$ lb/h.p.), good cooling and smoothness. However, its fuel consumption was high and its oil consumption very high, so most other manufacturers continued to develop in-line, vee and other non-rotary configurations. Many famous firms from the car and motorcycle world – Anzani, Panhard, Renault, Mercedes, Sunbeam, Beardmore and Austro-Daimler among them – became involved in the exciting advance in the air.

The First World War of course was responsible for the evolution of bigger, more powerful engines, primarily of the in-line and rotary types. Though the low weight of the latter gave them an advantage for fighter aircraft, the gyroscopic effects of the larger versions caused handling difficulties, and in due course they were replaced by the 'stationary' radial variety. By now Rolls-Royce had appeared on the aero-engine scene and soon assumed a leading role which they have maintained ever since. It must be admitted, though, that their first design, the Eagle, had certain affinities with Mercedes' earlier in-line six; two examples of this power unit happened to be in Britain at the start of the war and were sent to Royce who was being pressured by the government to build engines for military aircraft.

Post-war developments included the establishing of the air-cooled radial as a classic type, in both single- and multi-row forms, and the introduction of supercharging to maintain power output at high altitudes. The spread of engine size increased progressively to cope with everything from the smallest light aircraft to the biggest airliner, while refinements in design and advances in metallurgy brought impressive reductions in power/weight ratios. As an indication of what was being achieved in the latter respect less than thirty years after Orville Wright's first flight, the Rolls-Royce 'R' Schneider Trophy engine (one of the earlier V12s) of 1931 weighed just under 0·65 lb per horsepower.

In the early 1920s, Ricardo – mentioned earlier in connection with diesel-engine combustion – began to research the sleeve valve as a mechanically and volumetrically superior alternative to the poppet valve. This work formed the basis for Roy Fedden's renowned Bristol range of sleeve-valve engines (beginning with the Perseus in 1935) and later ones by Napier and Rolls-Royce. In spite of its superiority, particularly for military aircraft, the sleeve valve attracted surprisingly little attention outside Britain.

The era of the big piston-type aero-engine came to quite a sudden end in the mid 1940s with the advent of the gas turbine, though small piston units still abound for light aircraft. During those forty years or so, great strides had been made by the component and accessory makers as well as by the engine designers and builders. Before we look at some of these aspects in greater detail, though, there are two other technicalities of a general nature to consider. First of these is the fact that the two-stroke has had little attraction for aero-engine designers, most of whom considered that in the aircraft context its limitations outweighed its advantages. However, a technically noteworthy exception was the impressive Rolls-Royce sleeve-valve Crecy which appeared towards the end of the 1939–45 war but never went into service.

The failure of the diesel to compete significantly with the gasoline engine in heavier-than-air machines is more easily accounted for by its inferior power/weight ratio. It did have a certain appeal, however, because of the less flammable nature of its fuel, and several examples were built, notably by Bristol, Packard and Junkers; the latter's diesels were intended primarily for airships, in which the weight disadvantage was of relatively little significance. As a final comment on dirigibles, it should perhaps be mentioned that the early Zeppelins of the First World War were powered by converted car engines!

Carburettors and injection equipment

The Wright brothers' first engine had a type of surface carburettor, but the more powerful unit built for them in France by Bollée in 1908 featured a form of fuel injection into the inlet manifold. For those who have thought of gasoline injection as being a Second World War development, it is worth recording that even Bollée was apparently not the first to use this method since it was one of the revolutionary features of the Levasseur-designed Antoinette V8 constructed during the previous year. Float-controlled spray-type carburettors were preferred in those early days, however, because of their greater reliability, though the alternative of needle-valve control of the fuel was used by Gnôme on their Monosoupape rotary engines. No throttle was used with this arrangement, the engine being 'blipped' at idling by means of an ignition cut-out button.

By the end of the First World War most aero-engines were equipped with float-type carburettors roughly comparable with those used on vehicle engines. It is interesting that within a few years the American Stromberg company, which had not really entered the field until about 1918, had captured almost the entire world market for aircraft carburettors from other, older-established firms, and this situation continued into the 1930s. In its basic form, this variety of carburettor had two inherent disadvantages for aircraft duties: any violent manoeuvres – such as aerobatics – upset the mixture due to surging of the fuel, and icing-up of the throttle was liable to occur in certain combinations of atmospheric conditions and altitude. Because of their grip on the market, Stromberg were not under any duress to undertake a radical design and development programme, so they left these problems unsolved for over a decade.

When supercharging began to be adopted in the late 1920s, as a means of maintaining engine power at greater altitudes, another difficulty faced the carburettor manufacturer. Since air density diminishes significantly with altitude, while fuel delivery remains dependent on the air velocity through the venturi, the mixture becomes richer as an aircraft flies higher. This enrichment did not matter much when planes had relatively low 'ceilings', but supercharging enabled these to be extended upward considerably. Over-richness therefore became sufficiently pronounced to cause a loss of power as well as a reduction of range. The carburettor-makers' answer was automatic mixture control which progressively reduced the fuel delivery as altitude increased, by means of an aneroid capsule sensitive to atmospheric pressure (and hence air density). Devices of this type came into widespread use on military aircraft around 1934.

The next milestone was reached the following year when Chandler and Holly in the USA challenged Stromberg on their own ground by introducing floatless, pressurized carburettors which largely overcame the problems of altitude and accelerations in manoeuvring. In these carburettors, a diaphragm arrangement kept the fuel delivery constant, regardless of changes in the external situation. Such was the success of these rivals that Stromberg were forced into activity, and by 1938 they too had developed a pressure-type carburettor. This had the advantage of being virtually immune to throttle icing since the fuel outlet was downstream of the butterfly.

On this side of the Atlantic, though, the specialist carburettor manufacturers – including Claudel Hobson and SU in Britain –

remained faithful to float control but with a number of improvements. At that time the British instruments for the supercharged engines of the more powerful military aircraft were probably the finest of their kind. The amount of supercharge being applied by then to fighter and bomber engines had become so great in the search for power at altitude that there was a serious risk of over-boosting – and consequent mechanical or thermal failure – nearer the ground. This risk was obviated in the British carburettors by 'variable-datum boost control', usually of the three-stage type. In such a system, the pilot's throttle lever operated in a stepped gate giving maximum openings for take-off power, 'rated' or climbing power and cruising. At each stage the supercharge pressure was limited – through the butterfly opening – to the engine maker's prescribed figure by means of an interposed aneroid capsule similar to that in the automatic mixture control already mentioned. The gate position of the throttle lever determined the datum setting of the capsule which of course compensated for rising altitude by increasing the butterfly opening (up to the maximum available; any climb beyond that condition naturally caused the boost to diminish).

Serious work on gasoline injection, as an alternative to carburation, was initiated in the 1930s, primarily by Bosch in Germany but also by Bendix in the USA, following an earlier experimental programme by the military authorities. For all its improvement over earlier instruments, the pressure carburettor was not the complete answer for aero-engine fuelling because of some sensitivity to 'negative g' – the upward force encountered mainly as a centrifugal effect, when the aircraft's control column is pushed forward. Injection, of course, is not affected by this phenomenon, but it should be recorded that by the end of 1940 all SU carburettors on Rolls-Royce Merlin engines had been modified to overcome the 'negative g' sensitivity. Germany took an early lead in injection owing to two factors: carburettor development was less advanced there, and the country had already acquired a mass of injection experience with diesel engines. Inevitably, the outbreak of the Second World War caused a rapid acceleration of the development programme, and German military aircraft were being fitted with injection equipment at quite an early stage in hostilities.

On account of injection's operational superiority, as previously mentioned, Britain and the USA were compelled to introduce such systems too, in order to maintain parity. While most of these systems were the products of specialist companies, it is worth recording that Rolls-Royce actually evolved their own speed-density metering system at one stage for the Griffon engine which took over where the immortal Merlin left off.

On German military engines, the injection of the gasoline was effected either into the induction tracts downstream of the supercharger or directly into the cylinders, following diesel practice. Hence, the supercharger merely handled air. In contrast, British and American engineers favoured single-point injection into the supercharger intake, on the grounds of greater simplicity and the improved atomization of the fuel. Rolls-Royce used systems of this type produced by S U in Britain and Bendix in the USA, while Bristol's choice was the RAE (Royal Aircraft Establishment) equipment developed and produced by Hobson from original work done by Miss Shilling, one of the Establishment's most eminent experts. She, incidentally, was deeply involved also in the 'negative *g*' conquest mentioned earlier in connection with S U carburettors. The quality of the S U injection equipment is underlined by the fact that shortly after the war a US company, Simmonds, acquired the world rights – excluding the U K – for its manufacture.

It goes almost without saying that those complex carburettors and fuel-injection systems were developed to their high pitch of efficiency through very close collaboration between the specialist manufacturers and the engine companies. Installation details and bench and flight test programmes had to be worked out jointly to ensure optimum operational efficiency. In spite of the basic differences between the two methods of fuelling there was some 'spin-off' of knowledge from carburettors to injection, notably in respect of automatic mixture and boost control. In turn, when the gas turbine began to take over from the piston engine towards the end of the war, this acquired pool of combined knowledge was put to good use on the control side. As already indicated, too, some of the knowledge gained on aero-engine injection was applied in the automotive field.

Superchargers

Only one main type of supercharger – the centrifugal – has achieved widespread acceptance in the air, primarily because of its compactness, freedom from pressure pulsations and high efficiency in the conditions pertaining. However, two entirely different sub-types have been evolved more or less side by side: one is mechanically driven and the other is the turbocharger, driven by an exhaust-operated turbine.

Before we look at the development of supercharging, it is perhaps appropriate to point out that there is a fundamental difference – other than that of the driving system – between these two types of supercharger. Whereas the aircraft turbocharger has long been the province of the specialist manufacturer (although evolved in close collaboration with the engine companies), the mechanically driven variety has tended to be the direct responsibility of the areo-engineers. This fact might appear to exclude it from consideration in this book; however, apart from the affront to logic that would result from such exclusion, it is built from items which are very much the products of the specialists.

The first recorded work on the supercharging of aero-engines was in fact done at Britain's Royal Aircraft Factory (as it then was) at Farnborough in 1915. This programme quickly established the superiority of the centrifugal type, and a gear-driven version was an integral feature of the Factory's own 14-cylinder two-row radial engine which later became the Armstrong-Siddeley Jaguar. At about the same time as the Farnborough exercise, Auguste Rateau in France experimented with a turbocharger on aircraft engines, borrowing from the work of Büchi as referred to in the previous chapter. Farnborough took up Rateau's ideas in 1918 and carried on their investigations for several years, during which period both Bristol and Rolls-Royce designed their own turbochargers. In about 1925, though, the British Government decided to concentrate on mechanically driven superchargers as having fundamentally fewer problems. Thereafter virtually no attention was devoted to the turbocharger for aircraft engines in Britain until the late 1930s when Rolls-Royce had another look but decided against it in view of their wealth of experience with gear-driven units.

Progress in the latter this side of the Atlantic was thus continuous from the 1920s; apart from the already-mentioned Jaguar, Rolls-Royce produced a supercharged version of the Eagle in 1925 and of its successor the Kestrel in 1926. By the time the big piston engines were being ousted by the gas turbine in the mid 1940s, a line of development had been established leading from single-speed superchargers through the two-speed variety (to maintain power at still greater altitudes) to two-stage units with intercooling (for higher boost and maximum charge density) and even combinations of two impeller speeds and two stages of compression. The method of controlling the boost has already been covered in the previous sub-section.

As already indicated, design and development of these units was undertaken by the engine manufacturers. In the search for high efficiency, very careful attention had to be paid to such matters as inlet guide-vane and impeller form, and the gradual sectional change of the volute casing which collects the compressed mixture and feeds it to the induction tracts. This was where the specialists came in, since they had to produce substantial numbers of very precise light-alloy castings, steel forgings, etc., in addition to driving gears, bearings and – in the case of multi-speed superchargers – the clutches that affected the changes of gear ratio.

During the 1920s and 1930s supercharger development in Europe was on much the same lines as that in Britain. In the USA, though, the Air Force soon began to take an interest in the turbocharger. One reason for this appears to have been that the early attempts to produce mechanically driven units were in the hands of the component manufacturers, rather than the engine makers, and the results were not very satisfactory – presumably because the collaboration was not close enough. Another, more technical reason was that the military experts in the States were impressed with the fact that the power obtainable from an exhaust-driven turbine increases with altitude, so in theory enough boost should always be obtainable.

The mainstay of the US turbocharger programme was the General Electric Company. Such were the problems, though, that in spite of a fairly intensive programme, it was not until the late 1930s that satisfactory performance was obtained with reliability. Here of course the difficulties were two-fold because of the adjacent hot and cold sides of the turbocharger. In addition to the need for attaining good efficiency of the actual supercharger, as in the case of the mechanically driven variety, the turbine element had to make the best possible use of the very high-temperature exhaust gases. This meant the evolution of special high-temperature metals (high-alloy steels and, subsequently, non-ferrous alloys based on nickel) which did not lend themselves to easy production.

Increasing complexity resulting from technological progress was not confined to the mechanically driven supercharger. As an indication of this, it is perhaps appropriate to conclude this section with a reference to one of the last of the big piston aero-engines – the Wright Turbo-Compound, produced in the USA during the Second World War and developing nearly 3500 h.p. during take-off. This power plant had a conventional two-speed mechanical supercharger and no fewer than three exhaust-driven turbines. The power produced by these turbines was not used to drive additional superchargers; instead it was fed back into the crankshaft through hydraulic couplings and double-reduction gears – yet another task for the specialists!

Ignition equipment

Ignition on the Wright brothers' original flying machine was by means of a low-tension system of the type already discussed in connection with automotive engines. Those who followed close behind, however, relied mainly on coil ignition, and several years elapsed before the high-tension magneto had become sufficiently advanced to be trusted in the air. It is worth reiterating that much of the pioneering work here was done by Simms and Bosch, particularly the latter who established themselves before the 1914–18 war as the world's leading producer of magnetos for aero-engines. Rotary engines of course posed special problems here because no one fancied a bodily rotating magneto. A commutator and 'tight-wire' arrangement was therefore developed by the engine makers to enable the HT current to be transferred from a stationary magneto to the revolving cylinders.

Until about 1912 it was standard practice to have one sparking plug per cylinder, as in vehicle power units. As aircraft flew higher and faster, though, the hazards resulting from ignition failure of any kind became increasingly serious, so dual ignition systems (with two plugs per cylinder) began to appear. The first such system to achieve any fame – though it may have been preceded by others – was fitted to an Austro-Daimler aero-engine based on a car unit. Dual ignition soon became generally adopted for non-rotaries and has in fact been mandatory for many years on aircraft engines.

After about a decade of magnetos, the USA startled the flying world in 1917 by adopting Delco coil ignition on the 400 h.p. Liberty engine, which was a 'consortium' effort and not the product of a single company. Two official reasons were given – weight saving and the non-availability of suitable magnetos. The first reason is suspect on account of the need of a heavy battery, but the second one is valid since the Liberty's two banks of six cylinders were set at an included angle of 45 degrees, for low frontal area, instead of the usual 60 degrees. It could well be, too, that the choice of coil ignition was influenced by C. F. Kettering's return to it at General Motors for car engines. While the Liberty apparently had a good record for ignition reliability, no other

Robert Bosch was the pioneer of the integrated 'jerk pump' for injecting fuel into high-speed diesels; this early four-cylinder unit was built by his company in 1927. *Robert Bosch Ltd*

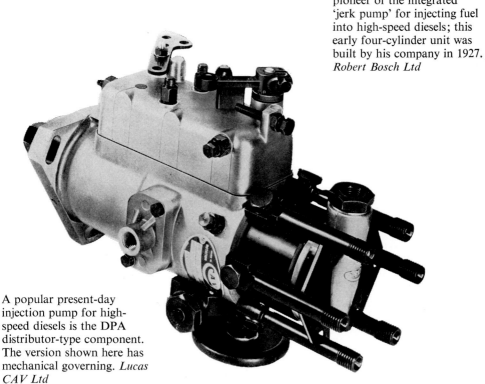

A popular present-day injection pump for high-speed diesels is the DPA distributor-type component. The version shown here has mechanical governing. *Lucas CAV Ltd*

Great precision is required in the
manufacture of diesel-engine
injection pumps; this illustration
shows the pumping-element barrels
being machined on a special-purpose
rotary transfer machine. *Lucas
Bryce Ltd*

Turbochargers have long been used to improve the performance of piston engines, in particular the aircraft and diesel varieties. Their development involves many hours of engine testing on the dynamometer. *Garrett AiResearch Ltd*

An advanced feature of the 1926 Triumph 14·9 h.p. car engine was automatic tensioning of the Coventry inverted-tooth chain driving the camshaft and the auxiliary shaft. *Renold Ltd*

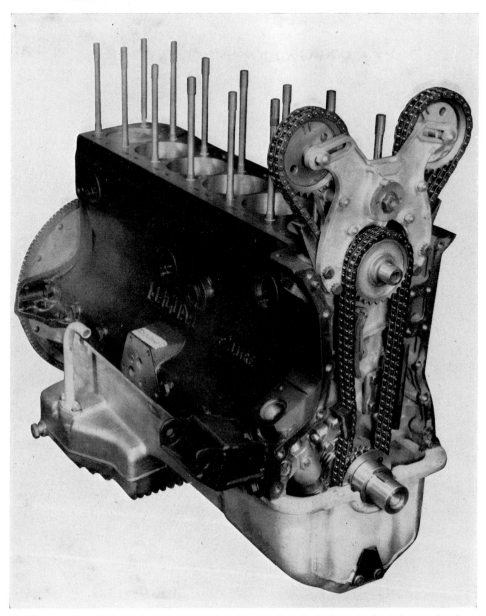

The Jaguar six-cylinder car engine has a two-stage roller-chain drive to its twin overhead camshafts; 0·375 in.-pitch duplex chain is used and an automatic hydraulic tensioner is incorporated in the lower stage. *Renold Ltd*

Opposite: In marked contrast to the Jaguar camshaft drive, this one for a 1971 Burmeister & Wain 28000 h.p. marine engine comprises three strands of 4·5 in.-pitch roller chain. A set of these chains weighs nearly 4 tons. *Renold Ltd*

The toothed belt has become a
common method of driving
overhead camshafts on car engines;
this application of such a drive is
to the V8 power unit of the
Porsche 928. *Uniroyal Ltd*

Ignition ancient and modern
– a Bosch high-tension
magneto of the early 1900s
and the latest Lucas electronic
system for the Jaguar V12
engine. *Robert Bosch Ltd
and Lucas Industries Ltd*

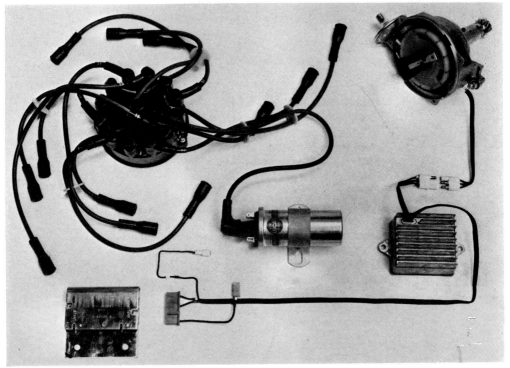

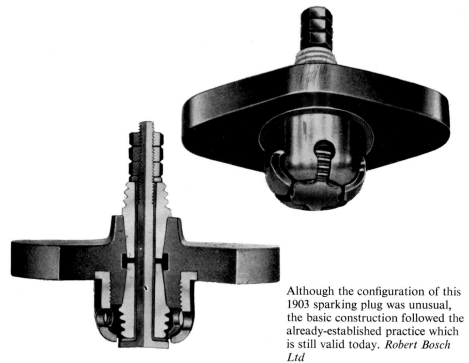

Although the configuration of this 1903 sparking plug was unusual, the basic construction followed the already-established practice which is still valid today. *Robert Bosch Ltd*

Below: Mica-insulated racing-car plugs, designed and manufactured before World War I by Kenelm Lee Guinness, racing driver and founder of the KLG company. *Smiths Industries Ltd, Motor Accessory Sales and Service Division*

Another old and new comparison – an early 18 mm sparking plug, embodying a tap for priming the cylinder with petrol, and a modern 10 mm plug for a Formula 1 racing engine. *Champion Sparking Plug Co. Ltd*

Four stages in the manufacture of the sintered ceramic insulator which has made such a big contribution to the efficiency and durability of the sparking plug of today. *Smiths Industries Ltd, Motor Accessory Sales and Service Division*

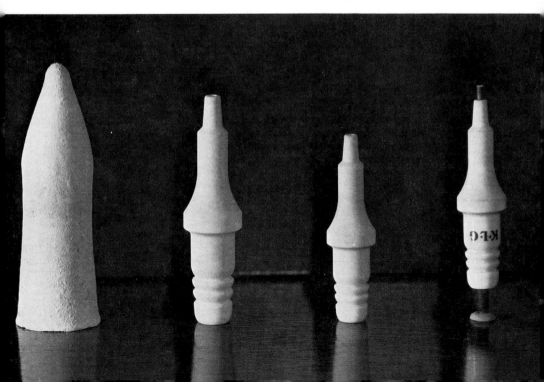

US engineer C. F. Kettering shows his first car-engine electric starter to Alfred P. Sloan, former President of General Motors; the photograph was taken in 1942, about 30 years after GM adopted this type of starter. *General Motors Ltd*

Two very different alternators – a 60/90 kVA aircraft generator as used in Concorde, and a 15 A car model with built-in rectification and control; the aircraft unit has an integrated constant-speed drive and oil cooling. *Lucas Aerospace Ltd and Lucas Industries Ltd*

Modern automotive radiator with a copper finned-tube matrix and brass top and bottom tanks; this type is robust and quite efficient, and lends itself to quantity production. *Covrad Ltd*

This fine example of the honeycomb radiator is from a Hawker Hurricane fighter of World War II; the cylindrical central portion is the integral oil-cooler element. *Serck Heat Transfer Ltd*

These illustrations of an early SU aero-engine carburettor of about 1920 and a 1941 four-choke unit, produced by the same maker for the Napier Sabre engine, underline the enormous progress made in little over 20 years. *SU Fuel Systems*

Because of the need for higher performance
and less sensitivity to acceleration effects, the
Rolls-Royce Merlin engines in some combat
aircraft were fitted from 1943 with this fuel-
injection unit in place of a carburettor.
SU Fuel Systems

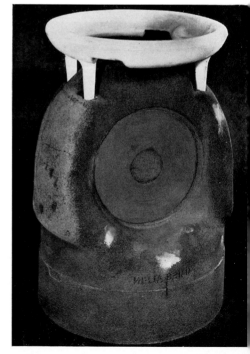

A salt core, for forming the oil-cooling gallery,
positioned on top of the internal sand core of a
medium-size diesel-engine piston.
Wellworthy Ltd

Final boring of insert-type bearing liners for large marine diesel engines. *The Glacier Metal Co. Ltd*

Big industrial and marine engines require complex cooling arrangements. Here, a system for a 20000 h.p. industrial power unit is being assembled at the factory; it features heat-exchangers for the coolant, lubricating oil and charge air. *Covrad Ltd*

manufacturer tagged along at that time, though coil systems are quite common today on small aircraft power units.

Until the early 1930s all magnetos were of the fixed-magnet/rotating-coil type. This arrangement has the advantage of not requiring very 'permanent' magnets, since stationary field coils can be used to excite the magnetic field, but is relatively heavy and has potential sources of mechanical and electrical unreliability in the revolving armature and the current-collecting slip-ring system. The electrical-equipment manufacturers therefore welcomed with open arms the development of new 'high-remanence' alloys of aluminium, nickel and cobalt which enabled them to produce magnets that retained their properties almost indefinitely. As a result they were able to go over to the mechanically and electrically more robust rotating-magnet type of magneto. Since these had stationary windings, the latter operated in a less stressful environment, and current collection was perfectly straightforward.

A few years later a third type of magneto was evolved and soon became quite widely used. It is known as the polar-inductor machine and its essential feature is that both the magnets and the windings are stationary. The inductor is a soft-iron member which rotates between the stationary components, producing a rotating magnetic field which is cut by the wires of the windings in the usual way. Although a little more complex than the rotating-magnet design, the polar-inductor magneto has the rotating-armature type's advantage of not depending on the permanence of the magnets.

The relatively small sparking voltage produced by a magneto at low rotational speeds made many earlier aircraft engines difficult to start, so means had to be found of improving the low-speed sparking character-istics. For larger engines the electrical folk added a 'booster' coil or magneto. The former acted exactly as in a coil-ignition system, while the latter was a small unit geared-up to run at several times engine speed during the starting procedure, so that it produced much higher-voltage sparks; as soon as the engine started, the booster coil or mag-neto was cut out of action.

For smaller engines, the propellers of which could be swung by hand, the answer lay in what is called an impulse coupling. This device, incorporated in the drive to the rotating-magnet member, holds that member back briefly and then allows it to flick through at a high speed to catch up with the drive. The acceleration takes place just at the in-stant the spark is needed, and the higher speed of course generates an adequate secondary voltage.

T.C.C.—D

Early history of the aircraft sparking plug is in general very similar to that for the automotive engine. Again mica replaced porcelain as the insulator material before the First World War, and its general employment continued until the 1930s when trouble began to be experienced through attack by the high lead content of the 87-octane and 100-octane aviation fuels of those days. However, mica remained in use for plugs in engines operating on 80-octane lead-free gasoline, and this situation carried on even after the Second World War. Prior to this war, though, the aero-engine plug had become considerably more complex than the vehicle type, primarily because the widespread adoption of radio in aircraft necessitated 'screening' of the plugs and harness to obviate interference.

It is appropriate at this juncture to look briefly at the operational differences between aircraft and vehicle plugs. The former have, on the face of it, an easier life in that conditions are more sustained – there are not such frequent variations of speed and load. On the other hand, take-off specific power outputs (and therefore temperatures) are considerable, and any piston or valve failure through pre-ignition is likely to be catastrophic. Consequently, the aero-engine plug has to be basically a higher-grade product than the automotive equivalent. For this reason it had become common practice in the UK before the 1939–45 war to use precious metals – such as platinum and its alloys – for the electrodes, and to incorporate more than one earth electrode. The platinum electrode was in fact developed by KLG in the 1930s, and further improvement came in the 1950s with the adoption of iridium by KLG and Lodge, another famous British plug firm.

When improved insulators became essential for engines running on high-lead fuels, as already indicated, various alternatives were investigated. Aluminium oxide (alumina), a very hard material of good thermal and mechanical strength, had shown considerable promise as far back as about 1914 but in pure form it presented major production difficulties. The first firm to come up with the answer to these was Siemens, in Germany, during the mid 1930s. They adopted the sintering technique: powdered alumina was formed into a blank which was then given the desired shape before being fired at high temperature to fuse into a solid mass.

Sintering not only enabled insulators to be produced economically in quantity but it also allowed other ingredients to be added to the alumina to improve the properties and facilitate manufacture. Because of the resulting interest in alumina, AC in the United States soon took out a licence for the Siemens material, while Bosch developed a variant which they in turn licensed to KLG in about 1936. A year or two later Lodge produced their own successful insulators of the same basic material as well as evolving improved manufacturing methods. (It is worth recording that Lodge later joined the Smiths Industries group of which KLG had been a member company since the late 1920s.)

By the outbreak of the Second World War, the alumina insulator had become fully established for high-output aero-engines. During the consolidation period and subsequently, various developments have taken place in respect of materials and production methods. One such method is to mix the powdered alumina with a binding agent and to extrude it in rod form for shaping by turning or form-grinding before firing. Another is the wet-pressing method in which the powder is moistened and pressed in a sac to form a blank; this is then shaped and fired as before.

For several years, in order to speed production by reducing the number of stages, AC moulded the powder directly to the final shape but this scheme was eventually abandoned because of the complexity and high cost of the steel moulds which had a relatively short life owing to the abrasive nature of the alumina. Yet another method, intended primarily for high-volume manufacture, was pioneered by Champion in the USA during the 1950s and has since become widely adopted. It is known as dry pressing and involves adding a wax to the powder to make it cohere when pressed in a mould of simple shape. The elimination of the moistening speeds-up production, the shaping and firing being as already described.

Because dry pressing produces a less strong insulator than can be achieved with extrusion (the strongest) or wet pressing, it is of interest that KLG have never adopted it for aircraft duties, though other manufacturers have done so. A feature of all these processes, incidentally, is that the shaped 'raw' insulator has to be nearly 20 per cent oversize in all dimensions to allow for the shrinkage that occurs in firing.

Starters and generators*

All the early aircraft engines were started by the simple though rather hazardous expedient of 'swinging the propeller', so the specialist

* To avoid undue fragmentation, this section covers equipment for gas turbines as well as piston engines.

companies were not involved. Even when engines grew larger, and power outputs reached 200 h.p. and more, hand starting by this method was still practicable because of the high leverage exerted from the tip of a propeller blade. The time arrived, however, when the task became beyond the compass of even several men – who ran the risk of getting in one another's way – so something better had to be found.

In the early 1920s, therefore, a man called Hucks invented a powered starter that was quickly taken up by the Royal Air Force where it remained in service for many years. The Hucks starter consisted of a wheeled chassis (usually derived from a car) carrying a small gasoline engine with its flywheel, clutch and gearbox, the latter driving a long shaft. On the remote end of the latter (the height and angle of which was adjustable) was a coupling that engaged in helical jaws on the hub of the propeller. The operating drill was to run the starting engine up to speed and then engage the clutch to rotate the aircraft's power unit; the combination of engine output and the energy stored in the flywheel was sufficient to spin the propeller quite effectively. When the main engine fired and over-ran the shaft of the starter, the coupling was thrown out of engagement by the helical jaws.

The Hucks starter was certainly an advance but it was a cumbersome piece of equipment, so a fair amount of time was needed to start a number of engines if only one chassis was available. As a result, interest in Britain began to turn by about 1930 to the use of electric starting for the bigger aero-engines – a method that has since become widely adopted. However, before we consider it more closely, mention should be made of another link in the chain of engine-starting history.

Shortly after the appearance of the Hucks starter, the inertia starter began to be adopted in the USA for both military and civil duties. Inertia starters were built by various ancillary-equipment manufacturers and formed part of the engine installation. They comprised two of the principal ingredients of the Hucks – a flywheel and a clutch – but the former was 'wound up' not by a secondary engine but by manpower: it was rotated by two detachable crank-handles, one at each side on the aircraft behind the engine, and these handles were manually operated. As with the Hucks starter, the clutch was engaged when the flywheel had been taken up to speed.

Because of the limited space available and the need to minimize weight, the flywheel could not be very big so it had to be spun fast to acquire sufficient energy to turn the engine effectively. This took time and also was quite hard work for the winders. Even so, the inertia starter was a sound item of equipment and continued in regular service until well into the Second World War. It is still encountered today but usually in a modernized form in which the flywheel is run up by an electric motor.

In this guise the inertia starter is of course one of the types of electric starting mentioned earlier, the other being the direct-cranking motor which operates on the end of the crankshaft. The inertia device has the advantage that, since the power requirement is relatively low and is absorbed over a significant period of time, the aircraft's own battery can be used to operate it. In contrast, the very heavy electrical drain of directly cranking a large piston engine over compression, and against the considerable oil drag, usually involves an external power source which has to be moved from one aircraft to another (this power source was originally a big trolley-mounted accumulator, but self-contained generating sets are now in common use). On the other hand, the direct-cranking starter is less complex, generally lighter and occupies a smaller volume.

For both these types of electric starter the motor manufacturers have done an excellent job in their endeavours to get a quart out of a pint pot, bearing in mind the ever-present requirements of low weight and bulk. With the advent of the gas-turbine engine their task became both easier and harder. Running-up the rotating assembly of a turbine unit to its idling speed is analogous to accelerating a flywheel – the effort is not very high and the time is quite long. However, the minimum self-sustaining speed of an aircraft gas turbine is likely to be 10000 rev./min. or more, while gearing-up the motor to slow it down would increase its required torque; the manufacturers have therefore had to develop motors capable of operating efficiently over a very wide speed range.

Interest in non-electric starters revived during the 1939–45 war, when the urgent need arose for systems that would give combat aircraft, particularly interceptors, the quickest possible start without dependence on trolley-accumulators which might not be available or could go unserviceable. The first such system was the British cartridge starter invented by Koffman during the late 1930s and often called after him though it was subsequently manufactured by Plessey. It was of course for starting big piston engines and, in effect, comprised a large pistol mounted on the engine; when the cartridge was fired, the rapidly expanding gases of the explosion operated on a piston which gave a hefty kick to the crankshaft, taking it through about 270 degrees.

Although the cartridge starter was mechanically simple, reliable and reasonably effective, it became redundant when the aircraft gas turbine replaced the big piston units for military duties. Lucas therefore evolved the Koffman's gas-turbine equivalent – the combustion starter. This was a bigger device and burnt cordite at a relatively slow rate; the expanding gases from the burning of the cordite were delivered through nozzles on to a turbine rotor which in turn revolved the shaft of the main engine. One charge of cordite lasted for upward of half a minute – ample to run the engine up to its self-sustaining speed.

Lucas's next stage, about 1955, was to replace the cordite charge by a combustion chamber fed with a monofuel propellant (so-called because it contains its own oxygen for burning); propyl nitrate was the most generally used monofuel. However, both these combustion starters required fairly frequent servicing because their efficiency soon fell off through the deposition of carbon in the nozzles and on the guide and turbine vanes of the starter.

A short-term answer to this operational disadvantage was the low-pressure air starter. This had a turbine arrangement similar to that of the combustion type, but the gases from burning cordite or propyl nitrate were replaced by air at about 500 lb./sq. in. (34·5 bars) pressure. In the early 1960s, though, Lucas developed the latest and most sophisticated of the line of starting systems for military gas-turbine aero-engines. It is based on a small gas turbine, electrically started and burning the same fuel as the main engine; this auxiliary unit is run-up to speed and then its power turbine is clutched-in to rotate the main turbine's shaft.

This gas-turbine starter, which was designed specially for the Harrier jump-jet aircraft, can perform other important functions beyond merely starting the main engine. The installation includes an electric generator, a hydraulic pump and a pneumatic pump; these can be driven on the ground by the auxiliary turbine to enable the aircraft's services to be checked-out without the need (and high fuel consumption) of running the main engine.

Early aircraft had no need of electrical generators. After the first year or two they all featured magneto ignition, and they had no lights or other equipment operated by electricity. Towards the end of the First World War, though, night-bombers came on the scene, so some form of interior lighting became necessary for the navigators. Soon radio was fitted to aircraft, and from then forward the electrical equipment has proliferated to become a major part of a modern aircraft's ancillaries.

The first generators were little DC machines mounted externally on the wing struts in streamlined casings and wind-driven by a propeller – not quite power for nothing, because of the drag, but a light, cheap and simple means of obtaining the limited amount of electrical energy needed at that time. As current requirements increased, though, the drag of the correspondingly expanding dynamo became excessive, so it had to be brought in with the engine and driven by it.

These engine-driven generators continued to be solely of the DC type until about 1940, when the alternator's advantages of robustness and higher speed capability led to its introduction and increasing adoption for aircraft duties. Although AC generation is the more common today, the DC generator has by no means faded entirely from the scene. It still has numerous applications, though these days it usually operates at 112 volts, not the 12 or 24 of earlier machines; one of the most recent developments in the DC field is the combined starter-generator for light aircraft, as a means of saving weight and bulk. Back to Kettering, with a vengeance!

The latest families of alternators operate at an even higher voltage than do the DC generators – generally 208 volts. Standard frequency too is high at 400 cycles a second, but in some applications it varies with engine speed because of the direct drive. Where the equipment to be operated requires a constant frequency, this has until recently been achieved by hydromechanical drives giving a steady output speed regardless of the input speed from the engine. Electronically controlled constant-frequency drives are now beginning to appear, though, and could well become standard in the not too distant future.

On both types of generator, the leading specialist companies – Lucas Industries, AC-Delco and others – have worked hard on improving output/weight/volume ratios and reliability, and on the development of electronic control systems and (for alternators) rectifiers of a basically similar type to those used on cars but far more sophisticated. It should be appreciated, however, that a far wider range of machines has to be built for the aircraft industry than the automotive. In the latter, a spread from about 75 watts to 750 watts covers everything from motorcycles to heavy commercial vehicles; while a light aeroplane may not need more than 150 watts, a modern jumbo-jet could have as many as four 120-kilowatt alternators to meet all its electrical needs.

Even alternators as big as these are still driven by the main engines. However, some large aircraft today, both civil and military, are equipped with a separate generator driven by an auxiliary engine; this generator is normally used on the ground for checking-out purposes, although it can be used in the air if necessary. Finally, windmill-type generators – with remarkably efficient fans and having an output of up to 25 kilowatts – have been developed for lowering into the airstream in an emergency. The wheel has indeed turned full circle, but it has picked up quite a lot of technology on the way!

Cooling media

The basic cooling of air-cooled aircraft engines is of course primarily the concern of the engine manufacturer, although he may well have to rely on the specialist producer of castings or forgings to provide the desired shape and finning for adequate heat dissipation. In the case of the earlier liquid-cooled power units, the primary onus was again on the engine company since each cylinder had its own water jacket, often fabricated and maybe even of copper for maximum conductivity and hence additional cooling by convection. Then, when aluminium monobloc construction was introduced in 1915 by Hispano-Suiza – on a V8 engine designed by Marc Birkigt – it became up to the foundry firms to comply with the designer's requirements for the water passages.

With liquid cooling, though, there is the secondary stage of transferring the engine's waste heat from the coolant to the air. As in the case of vehicle engines, the transfer medium is a heat-exchanger or radiator. At this point it is appropriate to interject that the heat transfer within the engine is normally to water (or a higher-boiling-point fluid such as ethylene glycol in some later power units) in the liquid phase, assuming no local vaporization occurs. However, some experiments were undertaken in the late 1920s by Rolls-Royce and the Royal Air Force on steam cooling, to reduce the amount of water carried. This work did not get very far, principally because in fighter aerobatics the steam and water tended to change places, with hardly ideal results!

Until the early 1930s, aero-engine radiators were of the conventional copper/brass honeycomb type; they were made by specialist companies who were not usually those producing automotive radiators which by then were already of different construction. These honeycomb assemblies were not only heavy but bulky, therefore doing much to offset the lower drag of the engine itself than that of an air-cooled unit. The first radical improvement here stemmed from work done at the RAE, Farnborough, by Meredith in 1935. He found that the energy lost in getting air through the radiator could be largely recovered by ducting the radiator in such a way that the energy added to the air from the coolant resulted in some useful thrust.

The ducted-radiator principle was eagerly taken up by Rolls-Royce who at that time were extremely interested in reducing cooling-system drag. They decided too that the honeycomb radiator had outlived its usefulness because its low air/water metal-exposure ratio of about 1 to 1 meant that a large amount of water had to be carried in the cooling system. As already mentioned, this point of cooling technology had been realized years earlier in the vehicle industry. R R therefore collaborated with Morris Radiators in 1937 to develop a secondary-surface radiator which had more than five times as much surface exposed to air as to coolant. As might be expected, this radiator resembled the latest automotive types, though its tubes were of deeper section.

Due to the much-improved design, it was found possible to reduce the frontal area of a Merlin radiator by about 10 per cent and the weight of coolant by no less than 40 per cent. Cost was reduced significantly too although it was not a prime consideration in this context. Then in collaboration with another specialist company, Marston-Excelsior, Rolls-Royce helped to pioneer a further cooling advance – the aluminium-alloy radiator, again of the secondary-surface type. The version adopted for the Merlin engines was actually 110 lb (50 kg) lighter than the original copper-tubed honeycomb assembly. On the outbreak of the Second World War, when quantity production became essential, the big British automotive heat-exchanger companies took over the manufacture of radiators for aircraft also.

So far we have been considering only radiators for liquid-cooled aero-engines. However, there is another important heat-exchanger application – the cooling of the lubricating oil. This is common to liquid- and air-cooled power units of most configurations but is especially critical in the case of the radial air-cooled type because of the very compact crankcase which provides little direct cooling for the oil.

As such engines became increasingly powerful, notably during the 1939–45 war, they needed more and more cooling of the oil to keep not only it but their own internal temperatures down to safe levels in combat conditions. According to a friend who was closely involved at the time, there was actually a widely held view among heat-exchange experts that these engines were primarily oil-cooled rather than air-cooled! Be that as it may, a lot of work was done then on oil cooling, and the outcome (in which a dominant role was played by Serck, another leading British heat-exchanger manufacturer) was the radial-flow oil cooler which marked an important advance in this aspect of engine technology.

These radial-flow coolers were eventually made of aluminium and other light alloys, for minimum weight, and were assembled by purely mechanical means – without any soldering or brazing – to ensure the highest standard of reliability. It is worth mentioning that the design was subsequently developed (again largely by Serck, although other companies were concerned also) into the high-pressure oil/fuel heat-exchanger commonly used on modern jet aero-engines, including those of our supersonic airliners.

No survey of aero-engine cooling would be complete without a reference to the internally cooled exhaust valve, since this has had a marked influence on the power outputs that could be obtained with reliability. The desirability of assisting the transfer of heat from the valve head to the stem (and thence to the guide and cylinder head) became appreciated during the First World War, and not very successful experiments were carried out at the Royal Aircraft Factory with valves having hollow heads and stems partially filled with mercury. The idea was that the mercury would be shuttled to and fro by the reciprocating motion of the valve, taking heat from the head and giving it up to the stem. Water was tried as an alternative, but the resulting high internal pressures led to leakage problems.

One of the engineers involved in this work was an American, Sam Heron, who a few years later (after his return to the USA) hit on the idea of using metallic sodium within the valve as a coolant. The principle of partial filling and shuttling was unchanged because sodium, though a solid at ambient temperatures, has a low melting point. However, the metal not only has a higher thermal conductivity than mercury or water but presented no sealing problems through vaporization. As a result sodium cooling proved both effective and practicable, and soon became widely employed on high-output aero engines in the late 1920s and early 1930s; its use duly spread to racing car and motorcycle engines which had similar problems of heat dissipation. The valve manufacturers played a vital role here in developing special precision-forging techniques to produce the internal spaces to adequately consistent dimensions.

3 Industrial, marine and locomotive diesels

Introduction

Although these three types of engine operate under different conditions, there is so much commonality in their technology that it is both convenient and logical to consider them together. All three applications embrace low-speed, medium-speed and high-speed units – as the terms are normally understood – although locomotive duties do not call for such small and high-speed engines as are used for some industrial and marine purposes. In any case, the smaller models are so closely akin to their automotive brothers as to leave no still-uncovered areas of importance.

Industrial engines were first on the scene and were followed by marine and rail applications at around the turn of the century. Right from the early days, the compression-ignition principle was favoured because of its operating characteristics, robustness and low fuel consumption; the inferior power/weight ratio of the diesel did not matter in the circumstances. The first recorded installations of compression-ignition engines in vessels and locomotives were in fact both made in 1902, the rail-traction unit being one of Ackroyd Stuart's original designs.

During the years before the First World War, industrial and marine engines in particular advanced rapidly, power outputs and sizes increasing considerably. Since both are primarily constant-speed types, they developed on similar lines which soon diverged to some extent from those of locomotive units for which a substantial speed range was a major requirement. Earlier engines all operated on the four-stroke cycle; two-strokes began to appear around 1912, however, though their true potential was not appreciated for about another ten years. Nowadays the two-stroke system probably has the advantage, certainly in the case of the bigger power units.

For many years the large industrial/marine diesels followed steam practice in being of the slow-speed type with the familiar crosshead arrangement to relieve the piston of side-thrust. This layout of course makes for a tall engine and, as dimensions increased, the phrase 'cathedral engine' came into use to describe the monsters that were being produced even before the First World War. The term is still used today, and with good reason when one appreciates that a modern example, developing perhaps as much as 4000 h.p. *per cylinder*, requires a five-floor lift to enable all the servicing stations to be reached!

After the first decade of the century, the crosshead engine's traditional open-crankpit layout (which by then was necessitating quite complex structures with columns and tie-rods) began to give way to all-enclosing crankcases. Another major advance in the pre-war era was the arrival of turbocharging, through the already-mentioned work of Dr Büchi whose company, Sulzer, was and still is one of the leading builders of large marine diesels. However, his ideas were not exploited on any scale until the early 1920s.

That period from 1920 to 1930 was actually a very significant one in the industrial, marine and locomotive engine field. In addition to the progress on two-strokes and turbocharging, a different type of diesel was being rapidly developed. It was the so-called trunk-piston engine in which the piston is linked directly to the crankshaft by a connecting rod, as in automotive practice.

The trunk-piston arrangement was made practicable mainly by improvements in metallurgy and production techniques, and its elimination of the crosshead not only considerably reduced engine height but made greater speeds possible through the diminution of reciprocating mass. To put crankshaft speeds into perspective, crosshead (slow-speed) engines operate in the 90–160 rev./min. bracket whereas trunk-piston (medium-speed) units have a 400–700 rev./min. maximum. The ability to run faster naturally meant that the trunk-piston type could produce more power for a given piston displacement, and hence for a given occupied floor space. Moreover, the lower height greatly facilitated the installation of a relatively big diesel within the confines of a locomotive.

One might think from the foregoing that the trunk-piston layout has sufficient advantages to have ousted the crosshead industrial/marine engine many years ago. That this has not occurred is due to several factors. Because of its low speed the crosshead variety is still regarded as the more durable of the two, and it is also the more efficient in really large sizes, so its place remains secure where very high outputs are required from one engine. Conversely, the trunk-piston unit is superior farther down the size scale; however, even over the range where the two are in direct competition, the crosshead type has the benefit in marine applications of not requiring a reduction gear to bring down the output-shaft speed to a level consistent with good propeller performance.

Yet another major step forward in the technology of both the low-speed and the medium-speed diesel came in about 1923 when Robert Bosch's high-precision jerk pump became available for fuel injection, as was discussed in greater detail in Chapter 1. This pump was a big improvement over the existing rather primitive 'airless' injection systems that had replaced the air-blast arrangement of most early engines. It paved the way, too, for the high-speed diesel (running at 1200 rev./min. or more) which made its impact at the end of the 1920s for industrial, marine and locomotive work as well as in motor vehicles.

There is no doubt that, since those days, the rapid progress in automotive diesels has influenced the design of medium-speed and low-speed engines in these other fields, as well as the high-speed versions. An apparent exception to this generalization is turbocharging, regarding which the flow of expertise was initially in the other direction: the turbocharger was already quite highly developed for the bigger engines before it caught the imagination of the vehicle engineers. Even here though, there has been a substantial feedback from the automotive sides in respect of design, manufacture and control.

On rail traction, the number of diesel locomotives in service on the European mainland, in the USA and elsewhere had already become significant by the early 1920s, but the engines were all large (or as large as could be accommodated) and ran at relatively low speeds. The 1920–30 advances brought in higher speeds and turbocharging for increased power, and the progress has culminated in the twin-engine (each of 1000 h.p. or more) installations in modern high-speed main-line locomotives.

It is of interest, in view of the earlier reference to divergence, that some recent locomotive engines have been 'borrowed' from the industrial and marine areas. This has been so particularly in Britain where –

because of her long tradition in steam and her big coal production – railway diesels were virtually unknown until after the Second World War, except for shunting purposes. These shunting prime-movers, incidentally, have always been derived from stationary industrial units since only a limited speed range is necessary.

A fundamental disadvantage of the diesel for rail traction, in comparison with the steam engine, is its lack of starting torque. Owing to the difficulties of evolving transmission systems to overcome this deficiency, there has been a growing trend towards diesel-electric propulsion for locomotives on heavy passenger and freight duties. In such an arrangement the engine becomes in effect an industrial unit, driving a DC generator at constant speed; the generator supplies axle-coupled DC motors which have an inherently high starting torque.

With the advent of the high-speed diesel another type of railborne vehicle – the railcar – became a practicability. Railcar engines, which have to be coupled to quite complex transmissions, are closely analogous to automotive diesels and are usually turbocharged to achieve the minimum installed bulk for the desired performance, in the interest of passenger accommodation. Since the early 1930s, their power outputs have risen – without dimensional increases – from about 100 h.p. to 200 h.p. and more.

Pistons, rings and cylinder liners

Since the builders of crosshead engines produce their own pistons and associated components, this section is concerned with those for trunk-piston diesel engines having cylinder bores in excess of about 6 in. (152·5 mm). As was indicated in Chapter 1, though, some of the problems that have had to be overcome are common to the very large automotive diesels of the type used in earth-moving and constructional plant.

The design of aluminium pistons for these larger engines did not present any major thermal or mechanical difficulties until the widespread adoption of turbocharging – and the consequent substantial increases in internal pressures and temperatures – during the era after the Second World War. Not only had pistons to be made stronger but means had to be found of getting the greater heat input away from the crown to avoid overheating. Intensive research and development work by the piston manufacturers during the 1950s and early 1960s showed that oil-cooling of the underside of the crown could significantly reduce temperatures in the critical areas.

Spraying the oil on to the crown by means of jets from the main-bearing supply proved very effective where the oil pressure was sufficient to ensure adequate jet velocity. This arrangement is still quite widely used, though mainly on highly rated engines at the lower end of the size scale under consideration. For larger engines, however, it soon became evident that a more positive form of cooling was preferable, so the piston makers began to investigate the circulation of oil – again under pressure from the main lubrication system – through a gallery within the crown.

There was no problem in getting the oil to the crown, from the big end by way of the connecting rod, but how could the gallery be formed? Normal casting-in was impossible because the sand core could not readily be removed subsequently from the enclosed annular space. One possible answer was the two-piece piston, of which there were two variants: in one, the complete crown with ring belt was bolted on to close an open gallery formed in the top of the piston's lower half; in the other, only the ring belt was separate, being shrunk-on to close a gallery behind it.

These designs were practicable but at that time they were considered undesirably costly as well as adding appreciably to the weight of the piston – particularly in the case of the bolted arrangement. The same criticisms applied to the alternative of casting-in a cooling coil within the crown. Both the objections were overcome very ingeniously by Wellworthy in a technique that they patented in 1964. Their scheme, which has since become firmly established, was to cast-in the passage by using a special salt core; after the piston was cast around the core, this was dissolved away with hot water, leaving the desired gallery. Simple and most effective!

The soluble-salt-core technique was subsequently applied to a 'half-way house' cooling system evolved by Wellworthy for smaller pistons, as an improvement over the orthodox oil-jet arrangement. In this scheme, the gallery is not completely closed but changes to an open section on the thrust and non-thrust sides of the piston. Jets spray the oil into one of the open sections during the upper half of the stroke, and it flows into the closed ones, where heat extraction is enhanced by the 'cocktail shaker' effect. The heated oil drains away from the open sections during the lower half of the stroke, due to inertia forces.

Engine ratings are continuing to rise, through higher turbocharging and other advances, so the piston companies are not able to relax. For variable-duty engines of ultra-high output, the ideal solution is the variable-compression-ratio piston which would ensure easy starting and optimum efficiency under part-load running without the risk of thermal or mechanical overstressing at full load. Several experimental pistons of this type have been produced; in particular the BICERI (British Internal Combustion Engine Research Institute) design has shown considerable promise. However, no engine builder has yet decided to adopt such pistons in view of their inevitably high cost in relation to the expected gains.

The alternative is of course to design for the worst possible conditions, and it now looks as though the next stage for medium-speed engines will be the two-piece piston with a steel crown and an aluminium skirt. One such design has been evolved by Ricardo & Co. the world-famous engine consultants, in collaboration with Wellworthy, for a high-bmep research programme. This piston has the compression rings in the steel crown, closed-gallery cooling and forged aluminium-alloy bushes in the gudgeon-pin bosses for additional strength; it has proved satisfactory for sustained running at a bmep of around 500 lb/sq. in. (34·5 bars) – well above present-day levels.

The design of these large heavy-duty pistons is far removed today from the empirical business it was fifty years ago. Various methods of calculating piston stresses have been developed and are in general use by the manufacturers; they involve a complete analysis of the mechanical situation and of the temperature distribution through the piston, based on a knowledge of the heat input and combustion characteristics. This theoretical work can nowadays be accurately checked by sophisticated telemetry techniques for determining temperatures and loads.

Little need be said about piston rings, since the technology discussed in Chapter 1 is mainly relevant to larger-bore engines also. Cast iron is the standard material for all the rings in marine, industrial and locomotive power units, and use is made of chromium plating generally for bores up to about 16 in. (407 mm). As in other engine categories, better design and production methods have made possible a general reduction in the number of compression rings for these larger engines from four or five – in most cases – to only three, in the interest of lower internal friction.

Diesels of above 6 in. (152·5 mm) bore are commonly of the wet-liner type, in contrast with the smaller-bore automotive units which generally have dry liners. A side-effect of uprating, in the case of wet-liner engines, has been the increasing incidence of cavitation erosion, to which a brief reference was made in Chapter 1. This form of erosion occurs on the outside of the liners where entrapped air or bubbles of steam tend to gather between the cylinders; the closer the cylinder spacing the more severe the erosion can be. It is attributed to pressure fluctuations caused by vibration of the liners, this vibration being excited by the impacts the pistons make upon them in passing through top dead centre and at about mid-stroke. The pressures involved here frequently exceed 2000 lb/sq. in. (137·9 bars)!

Cavitation erosion has been tackled in two ways, palliative and pro-phylactic. Under the first heading come thickening the liner walls where space permits (to reduce the amplitude of the vibration) and giving the external surface a resistant coating. In earlier days, ceramic and alumin-ium coatings were tried, but electrodeposited chromium soon became more generally accepted. However, plating in this way is a slow and costly process, so increasing interest is now being shown in coating with chromium or a cheaper chromium alloy by the plasma-spray technique mentioned in Chapter 1 in connection with piston rings.

The other attack on cavitation erosion has been aimed at reducing the severity of the piston/liner impacts; in this respect, comprehension of the underlying dynamics has been greatly assisted recently by exten-sive computer studies of piston movement. In Britain, Associated Engineering have been in the forefront of this research, both at the Group's R&D headquarters and at Wellworthy. Preventive methods developed by the piston makers, working closely with the engine firms, include offsetting the gudgeon pin, improving the piston skirt form so as to minimize running clearances, and providing oil-damping at the skirt bearing areas. The last-mentioned technique was developed by the Admiralty Research Laboratory in collaboration with Wellworthy. In addition to reducing erosion, these improvements have also been found to have a beneficial effect on noise emission, wear and internal frictional losses.

Bearings

In very general terms, the bearings of the smaller industrial and marine diesels, and of railcar units, involve a similar technology to that for the automotive engines already discussed. Consequently, there is no need of any further reference here. In the case of the bigger engines – the medium- and low-speed diesels – though, the basic conditions are different although most of the major bearings are still performing the same sort of functions. It is difficult to make any hard-and-fast distinction between the marine, industrial and locomotive fields in respect of bearings, for the reason given in the introduction to this chapter.

Medium-speed diesels operate in the 300–1800 rev./min. bracket of crankshaft speeds. Since most of them are four-strokes with trunk pistons (though the category does include a few two-strokes and crosshead examples), the layout and dynamics of the bearings are not far removed from those in the automotive field. However, the lower speeds brought less need for progress until the middle 1950s when turbocharging and other developments began to bring substantial increases in power-output ratings. Until then, the old practice of lining the housings directly with a thick layer of white metal had been satisfactory for crankshaft main and big-end bearings. This method actually continued for engines of earlier design until the mid 1960s; in more recently designed diesels, thin-wall inserts, with heavier-duty lining alloys, were usually specified from about 1955.

Though larger, of course, than their vehicle-engine equivalents, these thin-wall bearings followed the identical principle and were produced in the same way. Understandably, some of the bearing specialists already supplying the automotive thin-wall market – notably Glacier in Britain – extended their ranges accordingly. Initially the steel shells were lined with copper–lead or lead–bronze. Glacier had such confidence, though, in their recently developed reticular tin-aluminium material – from its excellent performance in vehicle engines – that in the late 1950s they introduced it in competition with the copper-based alloys for medium-speed diesels.

Since then, tin–aluminium bearings have gained wide acceptance, to the extent that they have captured around 40 per cent of the market in the UK, and the trend is still upward. Although Glacier also make bearings lined with both copper–lead and lead–bronze, the tin–aluminium type is their first preference for this duty, because of its freedom from corrosion and its compatibility with crankshaft journals. Other European bearing manufacturers have been obliged to follow suit, although they still retain a preference for the copper–based materials. At the present stage, tin–aluminium has made greater headway at the smaller end of the medium-speed size spectrum, but this could be because until recently 20 per cent tin–aluminium was not available in the overlay plated form; the latter helps to embed built-in foreign particles which are a great hazard in the larger engine.

In the many four-stroke engines of this type, the small-end bearings are often very heavily loaded. However, on alternate revolutions the inertia loading predominates over the gas loading, and this reversal of the load assists the formation of an oil film in the loaded half of the bearing; in the handful of two-stroke units, on the other hand, the loading is always unidirectional but its maximum value is lower. Because of these conditions the traditional bronze bushes are still quite widely employed for small-end duties on medium-speed engines, although the alternative of the thin-wall wrapped bush – lined with lead–bronze – is becoming more popular.

Most of the big low-speed diesels, being of the crosshead type, have a bearing 'problem area' not found in any other variety of reciprocating engine. The 'bottom-end' and main bearings have caused few traumas through the years; white metal is still the universal material for them and they have a very long life. Only recently, indeed, has there been any significant swing away from direct-lined bearings (actually most mains are of the thick-wall type), produced by the engine manufacturers, towards the thin-wall insert variety supplied by the bearing companies.

In contrast, though, the crosshead or 'top-end' bearings were long a source of trouble. This was because of the slow-speed oscillating motion and the unidirectional nature of the bearing load; as in the case of the small-end bearings of medium-speed two-stroke engines, there is no reversal of the loading to promote the oil film which contributes so much to a low rate of wear.

By the late 1950s the direct-lined bearing was clearly reaching the limit of its performance for top-end duties. The need for improvement then led to the adoption of the insert technique which of course also brought with it better control during manufacture and the easier replacement of worn or failing bearings. Glacier pioneered the fitting of insert-type top-end bearings, and tests were carried out by most of the leading engine builders around 1962. Following successful trials, Doxford and MAN soon adopted the insert liner for this duty; they were followed by Sulzer and Burmeister & Wain in 1970, and then shortly afterwards by Grandi Motori Trieste (formerly Fiat). Most of these engines were fitted with white-metal-lined inserts, and an improved material of this type was developed jointly by Glacier and their Japanese licensee, Daido Metal Co., in the early 1970s. Its grain structure is refined by the addition of cadmium and chromium, with the result that its fatigue strength is some 10 per cent higher than that of earlier alloys.

Realizing that even this advance would not be sufficient to cure the problem completely, Glacier were also evolving a 40 per cent tin–aluminium alloy, having characteristics very similar to those of white metal but with a higher fatigue strength at operating temperatures. Bearings lined with 40 per cent tin–aluminium were first made by Daido and were adopted in 1971 by Sulzer licensees in Japan. Because of the very arduous operating conditions of the top-end bearing, it is normal practice to plate both white-metal and tin–aluminium linings with a lead–tin overlay.

In our discussion on bearing technology, Glacier pointed out that, in spite of the much better performance of these latest components, the lining material is not the only consideration. They and other bearing makers are therefore currently investigating possible design improvements in collaboration with all the engine builders in the category under consideration.

Fuel-injection equipment

Although the function of this equipment on industrial, marine and locomotive diesels is identical with that on the automotive variety considered earlier, there is a fundamental difference between the two types. Except in the case of engines derived directly from vehicle units, the injection pumps are generally not integrated but individual, supplying one cylinder only. Also, the specialist firms manufacturing for this market have to cater for a much wider range of power outputs: whereas in the automotive field a span from 10 to 50 b.h.p. per cylinder covers almost all engines, industrial/marine requirements can be from 3 to 2500 b.h.p. per cylinder.

A further complication is the wide variety of engine types and duties – factors that prevent the injection-equipment manufacturers from producing a universally acceptable standard range of pumps and nozzles. Instead they have to make variations to suit engine builders' particular requirements while incorporating as many standard components as possible for economic reasons.

The above mention of 2500 b.h.p. per cylinder may appear to conflict with the earlier reference, in the introduction to this chapter, to power outputs as high as 4000 b.h.p. per cylinder. Currently no trunk-piston engines exceed the lower figure, the higher one applying to low-speed crosshead engines; the builders of these monsters usually make their own fuel-injection assemblies, while those that are bought-in are produced strictly to the engine company's specification.

Even in the early post-war years the injection firms would still be expected by the makers of medium-speed engines to advise on the combustion system as a whole. Nowadays, though, the power-unit designers have the necessary knowledge in that area, although joint consultation from an early stage is still essential. As in other spheres, the designing of pumps and nozzles has become much more scientific than it was, and a lot of time is saved today through the availability of computer programmes for injection system design.

Turning now to the actual hardware, the injection pumps for industrial, marine and locomotive diesels have all been of the 'jerk' type, as already described, since the demise of air-blast injection. Most of them, too, have metering control by the variable-spill method that was a feature of the original Bosch pumps. They are driven individually off the engine camshaft but at present are controlled from a common governor – a point that will be returned to later.

It is probably not generally appreciated that many medium and large diesels of the types considered here now have specific power outputs very much higher than are expected of automotive engines. Post-war developments, particularly in respect of turbocharging, have led to mean effective pressures of 225 lb/in.^{2} (15·5 bars) or more – nearly 50 per cent higher than in a typical turbocharged vehicle diesel.

The resulting progressive reductions in engine 'package size' for a given power output have posed problems for the fuel-injection specialists such as Bryce Berger (of the Lucas Industries group) in the UK and Bosch in Germany. Higher powers of course have meant greater fuel-delivery rates, while rising engine speeds have made the time factor of injection more critical, so higher operating pressures have become essential. In addition, long-term reliability and durability have had to be maintained. As a measure of the task here, marine and industrial engines are normally expected to run for 3000 to 6000 hours between injector overhauls, depending on running conditions, with a target life of up to 10000 hours. For injection pumps, a running period of 15000 to 20000 hours can be expected before any replacements become necessary.

By careful attention to design, materials and quality control, the equipment manufacturers have managed to keep pace with these toughening demands. As a case in point, Bryce Berger have succeeded in uprating some of their larger pumps by 50 per cent on output and 20 per cent on pressure capacity without any increase in the installed dimensions. Their maximum delivery pressure is now as high as 20000 lb/in.^{2} (1380 bars) for big trunk-piston engines – a far cry from the 5000–6000 lb/in.^{2} (345 to 415 bars) for the small marine/industrial diesels in which the fuel spray does not have far to carry within the cylinder. Even the largest automotive power unit does not require more than about 8000 lb/in.^{2} (550 bars).

Where really high pressures are used, the delivery pipes between injector pumps and nozzles become a critical factor. All the pipes of an engine should be of the same length, to equalize the trapped volumes of fuel, and the elasticity of the material must be limited to ensure that the pump behaviour is reflected accurately at the nozzle. The pipes are therefore of solid-drawn steel tube, with a bore of up to 7 mm for the biggest engines; they have to be of high-quality manufacture and corrosion resistant, and be carefully clamped on installation to minimize the risk of fatigue failure through vibration. According to the equipment makers, not all engine builders are as thorough as they should be in this last respect, which is why the pipework is still a significant source of trouble on large engines.

In ships and power stations there is a continuing trend towards unmanned engine rooms. These are fine in terms of operating economics but they magnify the seriousness of any fire hazard, such as fuel leakage from pipes. The result has been the development of sheathed high-pressure piping; any leakage from the inner pipe does not find its way into the engine room but flows inside the sheath to a catchment arrangement.

Nozzle design is in some ways more difficult than for vehicle diesels, not least because of the lower-grade fuels that are often burnt owing to their relative cheapness and ready availability. Some of these fuels – called 'heavy residuals' – have to be pre-heated to make them fluid enough to be fed to the engine. Such fuels tend to form heavy combustion deposits on the nozzles, with adverse effects on the latter's performance. One way of reducing these deposits is to cool the nozzles, and this practice is becoming increasingly adopted on high-output engines. The nozzle producers have therefore had to develop more complex designs incorporating passages for the coolant which is usually water but sometimes oil.

The choice of governor set-up for a marine or industrial diesel is made by the engine builder, from a range of proprietary units. Until about twenty years ago these were all based on steam-engine technology; they were of the flyball type, operating the individual rack-rods directly through a mechanical linkage. However, the big masses resulted in sluggish control, so something superior was sought. Hence the next stage was the hydraulic-servo unit of which the US Woodward company has supplied more than any other maker. Speed sensing in these governors is still by means of flyballs but the system is much smaller and lighter so its response is quicker. The flyballs actuate a hydraulic spool valve which feeds a servo cylinder connected to the pump rack-rods.

Hydraulic-servo governors are still popular today but here yet again electronics are moving in because of their precision and instantaneous response. In an electronic system, speed-sensing signals are obtained electromagnetically from the engine crankshaft and are fed to a control box which processes and amplifies them as necessary for onward transmission to the actuators; the latter can be electromagnetic, hydraulic or pneumatic. Electronic governing was being actively investigated at the time of writing by several specialist manufacturers and by engine builders, and it appears to be viable for engines of 1500 h.p. upward. The compactness of the actuating equipment involved could well lead to a simplification of the pump control system, with consequent improvement in the precision and rate of control.

4 Rotary engines

Gas turbines

Introduction

The ancestry of the gas turbine can be traced back as far as 1791 and the work of John Barber, while the idea of jet propulsion, using steam as the energy source, appears to have come to Hero in ancient Egypt. In the early 1900s the gas turbine reappeared in industrial usage – not very significantly at first but there was quite a lot of activity in this and other fields during the 1930s. Although the turbojet – the application of the gas turbine to jet propulsion – was invented in 1921, it lay fallow until 1936 when serious development of aircraft power units of that type began in both Britain and Germany, through the desire for higher engine performance, with minimal frontal area plus the ability to burn lower-grade fuels.

This indeed was the real beginning of the gas turbine's climb up the ladder to recognition as an alternative prime-mover to the piston engine. We in Britain tend to overlook the fact that the first-ever turbojet flight was that of Ohain's engine in a Heinkel 178 during August 1939, whereas the Whittle engine did not take the air, installed in a Gloster E28/39, until May 1941. Nevertheless, the Whittle design seems to have made a considerably greater impact on the British aero-engine industry than its German counterpart did in its own country.

It should be recorded, though, that during these early years of the aircraft gas turbine, Brown-Boveri (a well-known Swiss company) had been very busy in the industrial and locomotive fields. They were in fact the most successful of the firms responsible for the activity mentioned in the first paragraph: in about 1940 they had a 4-megawatt gas-turbine-powered generating set running at Neuchatel, and roughly a year later they produced a 2200 h.p. locomotive for the Swiss Federal Railways – both apparently 'world's firsts'.

During the Second World War virtually all the other gas-turbine research and development was, of course, concentrated on turbojets for aircraft. The technology has advanced rapidly ever since in this area, particularly in Britain and the USA, though inevitably the curve of progress has begun to flatten during the past few years. To demonstrate the advances made, thrusts of turbojets have gone up from the 850 lb of the Whittle W2B, through 6500 lb for the first production Rolls-Royce Avons to 40000 lb and even more for the latest big power units for airliners and military aircraft.

In the search for higher thrust and reduced frontal area, the original centrifugal compressors soon gave way to axial-flow designs incorporating multi-staging as a means of obtaining higher pressure ratios. Single-spool rotating assemblies were followed in due course by multi-spool arrangements in by-pass engines, leading up to advanced turbofan types as now made by Rolls-Royce in Britain and General Electric and Pratt & Whitney in the USA. Afterburning (or reheat) was introduced in the early post-war years as a means of obtaining augmented thrust for short periods, while materials and manufacturing techniques were improved to keep pace. Combustion systems too have been much improved through the years: the initial arrangement of multiple combustion 'cans', disposed around the longitudinal axis, has been superseded in aero-engines by part-annular and even fully annular chambers with a plurality of fuel-injection nozzles.

The turboprop, in which part of the turbine assembly drives a propeller or propellers (while the other drives the compressor in the normal manner), came on the aviation scene about 1945, Rolls-Royce being the pioneers. Since then, turboprops have been fitted to a number of outstanding civil aircraft, mainly of the shorter-range variety for which their characteristics were originally particularly well suited. Even in this field, though, the popularity of the turboprop slowly waned in favour of the turbofan which has now been developed far enough to show appreciable advantages, especially for the larger aircraft engaged on feeder-line work.

One cannot leave the aircraft gas turbine without reference to VTOL (vertical take-off and landing) jet engines, in respect of which Britain established an early substantial lead through the initiative of the now-associated Rolls-Royce and Bristol companies. There are two basic types of VTOL engine: one is the 'vectored thrust' unit, which converts from vertical to horizontal flight by deflection of the jet outlet nozzles, and the other is the lightweight lift jet; the latter can, if desired, be pivoted bodily to give forward instead of vertical thrust. Such engines can rightly be regarded as having added a new dimension to flying – one which, at the time of writing, has yet to be fully exploited.

Not surprisingly, the enormous success of the gas turbine in the air quickly led to an upsurge of interest in it for marine, industrial, loco-motive and automotive duties. One of the first non-aircraft applications was in 1946 when Britain's Royal Navy installed an aero-type unit experimentally in a high-speed craft. Then in 1951 a Shell diesel-electric tanker was partially converted to gas-turbine propulsion machinery, and some five years later a US-built Liberty ship was tried with a gas-turbine installation of no less than 6000 shaft horsepower.

These engines were of course the marine equivalent of turboprops, and numerous other examples appeared in the 1950s and subsequently. In some of the earlier big units the power turbine was driven from a separate gas generator of the free-piston type, but before long the more compact self-contained or integral system became standard – in general following aircraft practice. Although the fuel consumption of a non-jet gas turbine can be considerably improved by the incorporation of a heat exchanger (in which the waste heat in the exhaust gases is used to pre-heat the air prior to combustion), this equipment has so far been little used in marine installations because of its bulk and cost. Perhaps the world's diminishing resources of energy will necessitate a change of attitude here in the not too distant future.

Since the mid 1950s the gas turbine has come into shipboard use also for the driving of auxiliary plant – particularly for the generation of electricity – in competition with steam and diesel power units. These 'auxiliary' engines are of the type used in the industrial field, another in which the gas turbine has made a big impact in quite a short time. Many power stations and big industrial complexes, for instance, now have 'peak-lopping' or standby generating equipment driven by gas turbines, while some have base-load sets with outputs approaching 100 megawatts.

Two distinct varieties of gas-turbine prime-movers are used for electricity-generating purposes – those using a group of aero-derived nuits as gas generators for a separate power turbine, and those having the gas-generator and power-turbine sections as an integral assembly, as for the marine applications just mentioned; both can be made in large sizes. Since the former have a quick start-up ability (due to their ancestry, the relatively small size of their hardware and their low thermal capacity) they are the more attractive for standby duties. The large integral sets, on the other hand, are comparatively sluggish thermally and so are better suited to base-load operation. Because space is seldom critical in these installations, while low fuel consumption is clearly important, heat exchangers of the recuperative variety are sometimes incorporated.

The gas turbine has also been applied post-war to rail locomotive propulsion. British Rail took the initiative here, and both Brown-Boveri and Metropolitan-Vickers built gas-turbine locomotives of around 2500 h.p. for them soon after hostilities ceased. Although neither of these designs was particularly successful, interest continued and has born fruit in the high-speed trains that have recently appeared in a number of countries. Turboméca in France and K-H-D in Germany have been prominent in this field, using aircraft-type engines, and several comparable gas-turbine locomotives have recently appeared in the USA.

I have deliberately left the automotive gas turbine until last in this survey, because it has had the most chequered history and to date the least success. In the vehicle context, as well as the marine and industrial, this prime-mover has the environmental advantages of smoothness, a relatively clean exhaust and low noise level for a given output. It also has inherently good torque characteristics which allow simplification of the transmission – an especial 'plus' for heavy diesel-powered vehicles which often have to have two-speed final drives or dual-range gearboxes to give them enough ratios for efficient operation. On the other side of the coin, a vehicle engine undergoes more frequent and greater variations of speed and load than any other type. Since the conventional gas turbine responds comparatively slowly to such changes, designers have been faced with major problems of control, as well as difficulties in obtaining fuel consumptions comparable with those of equivalent conventional engines, diesels in particular.

In the 1940s and early 1950s, when everyone was very turbine-conscious, a number of major vehicle manufacturers began their own development programmes for automotive units – primarily with cars in mind. These companies included Rover and Austin in Britain, and Chrysler in the USA. The choice of car-size units (of 100 b.h.p. or so) may seem odd when one appreciates that with the gas turbine, generally speaking, the smaller the size the greater the difficulties. On the other hand, though, the car market had more glamour and possible prestige than the commercial-vehicle side; also, if the difficulties could be overcome at the bottom end of the size spectrum, larger sizes should be relatively easy.

The Austin project did not get very far, but Rover and Chrysler persevered. Rover were the first to have a practicable turbine-powered car on the road; this was in 1949, and in 1952 the same car inaugurated the world's land-speed record for such vehicles at over 150 mile/h. A subsequent power unit, installed in a sports-racing car specially built by BRM, performed well at the Le Mans twenty-four-hour race in the early 1960s. Meantime Chrysler had got to the stage of building a batch of 'production prototype' saloon cars for evaluation by selected customers, but actual production was quietly shelved shortly afterwards.

At much the same time, Ford and General Motors in the USA became interested in the gas turbine as a promising alternative to the diesel for long-distance coach and heavy-truck duties. After their Le Mans efforts, Rover concentrated successfully on the development and exploitation of the small gas turbine for industrial and aircraft-auxiliary use until the company became part of British Leyland in 1968. Then the interest was diverted into bigger 'commercial' units, in line with Ford and GM thinking. Although at one stage in the early 1970s it looked as though British Leyland really were about to produce a turbine truck, nothing had materialized at the time of writing. However, one or two smaller companies continue to be active in Britain and may yet cause BL to regret their 'low profile'. In the USA, on the other hand, GM are going into limited production of a unit for long-distance coaches, while Chrysler are pursuing their car-engine programme under the aegis of the Energy Research and Development Administration.

This relatively slow progress must be viewed in relation to commercial expediency as well as technical advantages. Operators would need to be convinced of overall economic as well as environmental advantages before they could consider replacing the automotive diesel by the gas turbine. Although the latter, at the same level of production, is regarded by many experts as inherently the cheaper (mainly because of its much smaller number of parts), the vast cost of scrapping 'traditional' engine tooling, and retooling for gas turbine manufacture, would inevitably be reflected in the selling price of the power unit.

Combustion systems for vehicle gas turbines have largely followed aircraft practice but the control aspect – as already indicated – has proved far from easy because of the particular operating conditions. The problem here is two-fold: conventional fixed-geometry set-ups give negligible engine braking when the 'throttle' is shut, plus a substantial time lag on reopening it before the torque builds up again. Both these characteristics, particularly the second, are clearly undesirable in a road vehicle, but they have been almost entirely eliminated by the use of variable-geometry turbine/nozzle groups. B L, Ford, Chrysler and G M have all played a part in the development of viable control systems (the last-named having evolved a variable mechanical coupling between the compressor and power turbines in the interest of better engine braking) and of course the manufacturers of blades, vanes and other specialized components have made them practicable.

The earlier vehicle installations did not incorporate heat-exchangers, for simplicity and minimal bulk. It soon became obvious though, that such equipment was going to be essential if the fuel consumption of vehicle gas turbines was to be brought down to acceptable levels, and the four engine companies have all done a lot of work on this aspect of the automotive turbine.

Turbine blades and nozzle guide vanes

In a multi-stage gas turbine, blades on the turbine rotor alternate with rings of stationary nozzles guide vanes (NGVs). These vanes direct the gases on to the turbine blades at such an angle that the latter derive the maximum rotational thrust; whereas the first NGV ring is receiving propulsive gas from the combustion chamber or chambers, subsequent rings redirect the gases emerging from the preceding blades. In terms of environment, blades and vanes are the most critical items in a turbine. Both types of component get very hot, particularly those nearest to the combustion area; while the guide vanes have to withstand rather higher temperatures than the blades they supply, at least they are static, whereas the blades are subjected also to the enormous stresses generated centrifugally by their revolving round quite a large axis at a very high speed.

Because of these especially arduous conditions and the need for extremes of precision in dimensions and form, turbine blades and NGVs pose exceptional problems in metallurgy and manufacture. Their design has always been the province of the turbine-engine builder, and in the early years the latter undertook all their production also. At the end of the 1940s, though, the engine makers began to entrust manufacture to sub-contractors who had the right sort of expertise and were prepared to specialize. That delegation has been extended through the years, and today a high proportion of gas-turbine blades and NGVs are produced by outside firms.

Throughout the western world there are less than a score of these 'super-specialists', deservedly so called because of their unique combination of know-how, production and testing equipment and concentrated liaison with the engine companies. The marketing director of one of the leading British blade and vane makers – AE Turbine Components – told me that the complexity of form and the metallurgical and production problems had blended into a way of life for him and his firm; not only was this something far beyond the scope of any normally competent component manufacturer but it had spoiled AETC for any other lesser activity.

The continuing nature of these metallurgical and manufacturing difficulties has stemmed from a single fact: the power available from a gas-turbine engine is a function of the temperature at which the actual turbine operates. Consequently much of the struggle to advance has been directed at raising the practicable turbine inlet temperature, and this has necessitated the development of ever more heat-resistant blade and NGV materials. By the natural order of things, the tougher these are the more difficult they become to form into the required shape.

Although the materials are primarily the concern of the metallurgical companies, the engine makers and the blade/vane specialists are deeply involved – the former in respect of service performances and the latter on the production side. The precipitation-hardened austenitic steel used for the forged 'hot-end' components on Sir Frank Whittle's W1 engine soon proved to have inadequate creep strength, the blades tending to elongate under sustained high stresses. In Britain that steel was therefore replaced by Nimonic 75, the first of an advancing series of nickel-base 'super-alloys' of that name. From the mid 1940s up to the present day, these Nimonic alloys (which are hardened by aluminium and titanium) have been the principal materials used in British and European aircraft jet engines for forged blades and guide vanes.

During the war years, the Germans were too short of nickel to employ it in this way for their jet engines. Consequently they had to stay with austenitic alloy steels and developed a titanium-hardened one of improved properties. At that time their blades and vanes were not forged but were fabricated from sheet and had to be air-cooled to keep the temperatures down to acceptable levels. In contrast, the earliest US gas turbines had cast components of cobalt-base alloys, but creep became a problem with these as operating conditions toughened, so a change had to be made to nickel-base materials the first of which were forged – as on this side of the Atlantic.

American hot-end technology did not long remain on the same path as that in Britain and Europe, however. By the early 1950s the US experts recognized the limited development potential of the nickel-base alloys which in any case were automatically becoming increasingly difficult to forge as their high-temperature creep resistance was improved. By switching from forging to high-precision investment casting (see later) the Americans not only simplified production but opened the door for the introduction of more complex alloys of considerably higher hot strength than had previously been practicable.

Being suspicious of the reliability of investment-cast blades and vanes, the European engine firms – including the British ones – persevered with their forged nickel-base alloys until the early 1960s. By then it had become clear that the wrought alloys had reached the end of their development road; hence there was no option, if competitiveness was to be maintained, but to follow the US lead and go over to cast components and more exotic alloys for engines of the highest performance. The specialist suppliers therefore had to master the new techniques, and set up the relevant facilities, in a very short time. Because of the doubts just mentioned, they were required to make NGVs before the more highly stressed rotating components.

Every coin has two sides and on the reverse side of the hot-strength penny is one word – corrosion. As yet another example of the cussedness of nature, compositional modifications to improve the temperature capability of an alloy also reduce the resistance to hot-corrosion, basically because of the necessary reduction in the chromium content.

To complicate the situation further, the corrosional environment for the aircraft gas turbine differs from that for the industrial type. The former breathes clean air and burns high-quality fuel in which the only significant contaminant is sulphur which forms corrosive sulphides during combustion. Industrial turbines (embracing marine and loco-motive duties also), on the other hand, consume much cruder fuels and frequently have to operate in contaminated atmospheres. Not only are their hot-end components attacked by other corrosive combustion products, as well as sulphides, but the engines are expected to operate for very long periods – sometimes as much as 100 000 hours – between overhauls.

One or two of the latest nickel-base cast alloys can meet both the strength and corrosion-resistance requirements of most industrial gas turbines – because their temperatures are lower than those in top-performance aero-engines – and so are coming into increasing use in that particular field. Where these alloys are not adequate, the present alternative is an aluminide protective coating produced by a diffusion process; until the arrival of the 'compromise' alloys just mentioned, these coatings were applied to almost all turbine blades and some NGVs. More complex coatings, involving chromium and other metals as well as aluminium, are being developed in the search for still higher resistance to corrosion.

Turning now to the actual production, the manufacturer has first to be able to melt the materials from which he will be making the blades and NGVs. For forged components the melting is necessary to produce the ingots for the forging presses, while for investment casting the metal has of course to be poured. Virtually all the super-alloys used for the hot-end parts for high-performance engines oxidize very easily when melted in an air environment, and the presence of oxide dross on the surface would cause defects in the ingots or castings. Consequently the melting has to be done in electric vacuum furnaces which consist of a crucible surrounded by an induction coil, all situated in a vacuum chamber.

Forging techniques have advanced to such an extent that it is now possible to produce the aerofoil portions of blades and NGVs so accurately that they do not require to be machined afterwards. Even where these precision-forging methods are used, though, machining of the root fixings or any tip shroud details continues to be necessary, so a real difficulty still arises from the propensity of the alloys employed to work-harden during that machining. The energy absorbed in work-hardening is released on subsequent heat-treatment, manifesting itself as distortion of the component.

Even quite minor distortion can bring the blade or NGV outside the very tight tolerances demanded by the drawing. At best the item then has to undergo costly and awkward reworking, and at worst it has to be scrapped. Consequently, the specialist manufacturers have to take all possible steps to minimize work-hardening; the cutting tools have to be geometrically correct and meticulously and frequently sharpened, while the mounting of the workpiece on the machine has to be ultra-rigid, as of course does the machine itself.

Work-hardening is not such a bugbear in the case of investment-cast components. Not only is considerably less machining normally required than on a wrought part, because of the greater precision of the casting process, but it can be effected by grinding techniques which, if properly carried out, do not transmit nearly so much energy into the workpiece as does the profile-milling, broaching or turning employed on the forged items. On the other hand, a casting inevitably contains small voids and/ or inclusions, the presence of some of which is difficult to detect by even the best of present-day non-destructive testing methods, especially in heavy sections. The defects are therefore rarely discovered until after the machining stage – an expensive way of finding out whether a component is to be used or scrapped!

For those unfamiliar with the process, investment casting begins with the preparation of an exact replica or pattern of the component, in a low-melting-point material. On this replica is built-up a multi-coat refractory 'shell' of zircon or alumina, bonded with an agent such as sodium silicate. Then the replica is extracted, as explained in the next paragraph, leaving a precisely shaped cavity into which the molten metal is poured to produce the casting.

Although this casting method was apparently known to the ancient Egyptians (who used wax for the patterns and burnt it away after moulding – hence the alternative name of 'lost-wax process'), it achieved very little prominence thereafter until adopted in the USA for hot-end turbine components. Initially the Americans made their patterns of frozen mercury but this had the disadvantages of high weight and cost as well as being a health hazard. They therefore reverted to wax, specially formulated for high strength, and soon introduced the production economy of melting it out in a high-temperature steam autoclave, from which it could be recovered, reprocessed and used again.

Because of the already-mentioned machining difficulties with the wrought alloys, the specialist companies began in the 1950s to adapt two experimental non-mechanical forming techniques to their particular requirements. AETC were among those to pioneer these methods which proved so successful that they are still in use today. One is electrochemical forming (ECF) which in principle is the reverse of electroplating. The component to be machined is made the anode in an electrolytic cell; the cathode, generally of a conductive but corrosion-resistant material such as stainless steel, has a working surface that is the negative of the shape to be produced.

When a heavy current is passed at low voltage from the anode to the cathode across the small gap between them, the workpiece is dissolved away until it mirrors the shape of the cathode. In this way, quite complex three-dimensional forms can be reproduced with reasonable accuracy. ECF can also be used to produce very deep holes that would be impossible by conventional drilling methods; such holes are a feature of the blades and NGVs of most modern gas turbines, principally for air-cooling and/or reduced weight but also for installing the temperature probes mentioned in the section on monitoring and control.

The other non-mechanical method is electrodischarge machining (EDM). Here again the workpiece is the anode while the cathode is a negative of the shape required, but instead of the electrolyte a dielectric fluid – generally a light oil – surrounds the electrodes. Passage of the current creates a spark from the cathode which removes particles of metal from the anode.

EDM is used to produce very fine holes, slots and other shapes beyond the scope of either normal machining or ECF. However, it has its disadvantages: it leaves a rougher finish on the component than do other methods, and on the immediate surface there is an unstable layer of recast metal resulting from the heat of the spark. Unless this layer is acceptable in terms of the surface stresses involved, or is accessible for subsequent removal, then EDM is not suitable.

These electrical techniques have recently been joined by an improved mechanical one called 'creep-feed grinding'; it is particularly suited to cast components from which relatively large amounts of material have to be removed. This requirement may seem strange in view of the earlier statement that investment casting minimizes the amount of machining. The reason is that the very tight tolerances (of the order of ± 0.0002 in. or 0.005 mm) on serration-type root fixings – as commonly employed for blades – cannot be achieved by casting. Hence, the root ends are cast as blanks and the serrations are ground on to them; for a large blade, as much as 0.5 in. (say 13 mm) has to be removed, so heavy cuts are necessary for quick production.

Creep-feed grinding is fundamentally different from conventional grinding. The latter involves the reciprocation of the workpiece beneath the grinding wheel which removes very little metal during each pass. In the creep-feed process, on the other hand, the workpiece is stationary;

the wheel, which is in continuous contact with the workpiece, takes a very deep cut – sometimes as heavy as the half-inch just mentioned – during its slow advance. The special machines used have to be very sturdy to cope with the high forces generated, so they are expensive items of equipment.

Responsibilities of the specialist manufacturer do not end with the final machining operation. Meticulous inspection and testing are necessary to ensure that the components will stand up to their arduous duties, and these activities involve a lot of sophisticated modern apparatus. Dimensional inspection is carried out by electronic and high-magnification optical methods, while a wide range of radiographic, ultrasonic and other non-destructive techniques are employed for testing. Only when the blades and NGVs have proved satisfactory on all counts, and are ready for despatch to the engine builder, is the task of the specialist completed.

As to the future, many turbine experts believe that the continuing search for higher efficiency, and therefore lower fuel consumption, will lead to the adoption of blades and NGVs of a ceramic material (such as silicon nitride) instead of metal. This is because ceramic components could operate safely at turbine inlet temperatures about 200°C above the highest super-alloy levels, so a significantly smaller engine would be needed for a given output. Although quite a lot of exploratory work has already been done on ceramics, their characteristics are so different from those of metals that reliable performance criteria have yet to be established. Once these exist, appropriate production techniques will then have to be developed by the manufacturers, so the era of the ceramic turbine looks to be ten years or more ahead.

Combustion and fuelling systems

The combustion system of a gas turbine comprises the actual combustion chamber or chambers and the associated equipment for delivering the fuel. That equipment in turn consists of the fuel pump, which feeds the appropriate quantity under the command of the control system (see next section of this chapter), and the injectors that spray the fuel into the chamber in a pattern and droplet size giving the best possible burning characteristics.

Although the manufacture of gas-turbine combustion chambers has always been the responsibility of the engine companies (who may of course contract-out the production to metal-working specialists), the *design* of these vital components requires the closest possible collaboration with the fuelling-equipment producers. These firms, of whom Lucas Aerospace in the UK are probably world leaders, have built-up impressive R&D facilities through the years and have acquired a fund of knowledge that is invaluable to the engine constructors.

The earlier aircraft gas turbines, and their derivatives for other duties, had a ring of combustion 'cans' – individual chambers running axially or obliquely from the compressor outlet ring to the turbine intake; each of these cans had its own injector arrangement. Continuous research led, however, to the development first of part-annular and then of fully annular chambers. The latter – of which only one is needed per engine, of course, are now featured on most modern GT power units of all types. Welded construction has always been employed, using heat-resistant alloy steels at first, and then, as operating temperatures rose, more exotic materials such as the nickel-base Nimonics and even ceramics.

Combustion-chamber form has to be evolved in conjunction with the fuel injectors, as a combined sub-system. Since the injectors can spray from the upstream end or from inboard or outboard of the chamber, as well as pointing either against or with the airflow, several configurational possibilities have to be considered for each basic design of the chamber itself. On top of that, the fuelling specialists have developed numerous varieties of injector – including those achieving atomization by fuel pressure alone (pressure jet) or with the utilization of air (airblast or air-assist), and capable of producing conical or fan-shape sprays – in the search for the most efficient burning.

The chamber/spray correlation work is done partly empirically (suck it and see!) and partly by various flow-visualization techniques. For the latter, full-size replicas are made of the proposed designs and are subjected to water-flow tests: in these, the flow pattern can be observed by injecting air bubbles into the water supply, upstream of the test unit, and shining a thin parallel beam of light across the appropriate section of the chamber. Under certain conditions it is possible to inject dye to simulate the fuel and, by means of special instrumentation, to assess the probable local air-to-fuel concentration and 'mixedness' in the actual combustor.

Much time and energy has been devoted to these aspects of gas-turbine technology by Lucas Aerospace at their Burnley research laboratories. Mindful, too, that combustion has to take place *in* something, and that this something must stand up to its task, the company has extended its studies to embrace both the materiology of combustion chambers and the fabricational techniques employed in their manufacture. Other firms on the fuelling side have made significant contributions, on both sides of the Atlantic, but none appears to have the same breadth as well as depth of involvement – and certainly not a higher competence.

Lucas have long been in the thick of things too regarding fuel pumps, their existing axial-piston unit being a development of a similar type of pump used on the first Whittle gas turbine. They have always remained faithful to this type of pump for gas-turbine duties, primarily because it gives surface and not line contact at its working interfaces. US companies, on the other hand, have preferred gear and (to a lesser extent) vane pumps; since failure of either of these varieties is never catastrophic, perhaps one can put down the American choice to an innate pessimism about reliability!

Certainly the fuel pump of a gas-turbine engine does not have an easy life, because it has to handle fuel which not only contains a certain amount of dirt but also has poor lubricating properties. In fact, the 'lubricity' of turbine fuels – especially those for aircraft – is significantly worse today than it was thirty years ago, owing to the latest refining techniques and processes necessary to meet current fuel specifications in other respects. To conquer these adverse conditions, the makers of axial-piston pumps have had to introduce special measures such as the use of carbon facings on the bores and the slipper pads that bear on the swashplate; and, of course, the precision of manufacture usually has to be very high.

In contrast with the practice in some other applications of axial-piston pumps, those for gas turbines usually have the pistons in a rotating body, the camplate being stationary. This arrangement makes it a simple matter to vary the fuel delivery by altering the angle of the plate and hence the stroke of the pump. The alternative to varying the pump delivery is to have a constant output and to recirculate the appropriate amount of fuel back to the pump input; recirculation means that more work is put into the fuel, and the consequent temperature rise could lead to troubles at high altitudes in the case of an aero-engine.

As the size and performance of gas turbines have increased, the pump makers have had to step up maximum delivery rates accordingly. Back in the days of those first Whittle engines, 500 gallons per hour (2273 litres/hour) was ample for each unit. Nowadays a typical figure would be 1200 gal./hour (5455 litres/hour) at 3000 rev./min. but, as engines are

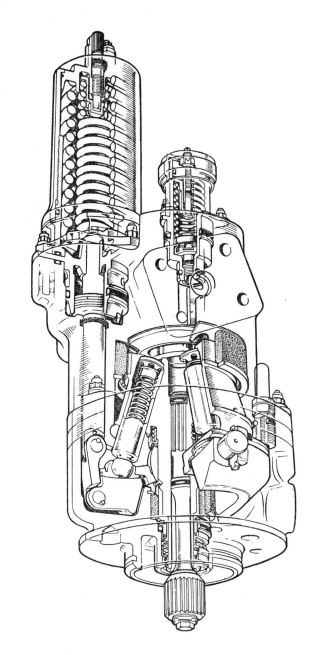

A Lucas Aerospace multi-plunger pump with variable delivery, of the type that supplies the fuel to many current aircraft gas-turbine engines

much more efficient today, this delivery would produce many times the thrust of the Whittle turbine. As an indication of what can be achieved with axial-piston pumps, units have been built for fuel flows as high as 6000 gal./hour (about 27 300 litres/hour) at 2000 rev./min. However, most of the pump manufacturers would prefer at the present stage to achieve such a delivery with two smaller pumps rather than one very big one.

Larger engines usually necessitate higher delivery pressures because of the longer 'carry' required of the nozzles. These higher pressures can readily be obtained at any rotational speed from axial-piston pumps which, because of their design, do not suffer from distortion or unduly high localized stresses. Looking to the future, Lucas Aerospace are developing a radial-piston pump of double-row configuration – in Bristol Hercules style. The rows are phased to balance the pressure and inertia loadings on the bearings, thus permitting the use of unusually high speeds without unduly rapid wear. Lucas themselves admit, rather deprecatingly, that this layout 'has prospects'.

Owing to the conservatism of the US gas-turbine industry, the ancillary-equipment firms across the Atlantic have contributed little to fuel-pump technology. As already indicated, they have concentrated mainly on the gear type but current designs are not very different from those of the 1950s – only bigger. To underline this slow progress, the only significant advance in gear pumps is now being developed in the UK: the new unit is of the internally meshing type which is quite common today for lubricating-oil pumps and the hydraulic systems of vehicle automatic transmissions. Advantages claimed over the conventional gear design in the turbine application are greater compactness in relation to output, lower stress levels on the teeth and reduced bearing loadings. Small wonder that the US engine companies are showing interest!

At the present stage, less development effort has been put into the centrifugal fuel pump than any of the types discussed previously, although such pumps are fitted to the Concorde supersonic airliner. At first sight the centrifugal pump may seem an excellent choice on account of its absence of rubbing contact and its consequent complete insensitivity to dirty or non-lubricating fuels; in addition, its manufacture involves less high-precision methods than does that of piston or gear pumps. However, it has a major operational disadvantage: it is not self-priming and so has a slow rate of pressure rise for starting purposes unless an additional priming pump is embodied in the fuelling system.

Finally, the Russians are notoriously secretive about many of the details of their aircraft and other gas turbines. It is therefore pleasant to report, in case any further proof is needed of Britain's major contribution in the area we are considering here, that most of the Russian engines have fuel systems deriving largely from the work of Lucas Aerospace, although this company regards them as being rather primitive in execution.

Monitoring and control

For a gas-turbine engine to operate efficiently and without the risk of severe damage, it has to be accurately controlled, particularly in relation to its rotational speed and its internal temperatures and pressures at and near its maximum putput. Overspeeding means excessive centrifugal loadings, and hence the risk of mechanical disaster, while overheating can have a number of ill effects including creep of the turbine blades and consequent vanishing running clearances. The control aspect is most critical in the case of aircraft power units, because of the special operating conditions and the dangers to life attendant on any failure, but it is important too in all other gas-turbine applications.

Great strides have been made in the development of 'engine management' systems and equipment since the days of the first aircraft gas-turbine engines, and all credit must go to the handful of specialist companies who have worked very closely with the engine manufacturers in this area of technology. In Britain, where so much of the pioneering has been done, Lucas Aerospace and the aviation divisions of Plessey, Dowty, Smiths Industries and Ultra have all made substantial contributions.

The primary control of a gas-turbine engine is on the fuel delivery, since this affects both speed and temperature. Aircraft power units have to operate from ground level up to considerable heights and so, as has already been mentioned in connection with carburettors for piston engines, the fuel flow must be reduced as the air density falls with increasing altitude. To simplify the pilot's task, even the earliest aero-turbines had a barometric-pressure-sensitive element between the pilot's 'throttle' lever and the delivery control on the fuel pump, to avoid any altitude restriction on the lever travel.

By the mid 1940s, the increasingly powerful engines were posing stability and reliability problems not previously encountered. The real need now was for some sort of acceleration control to deal with transient

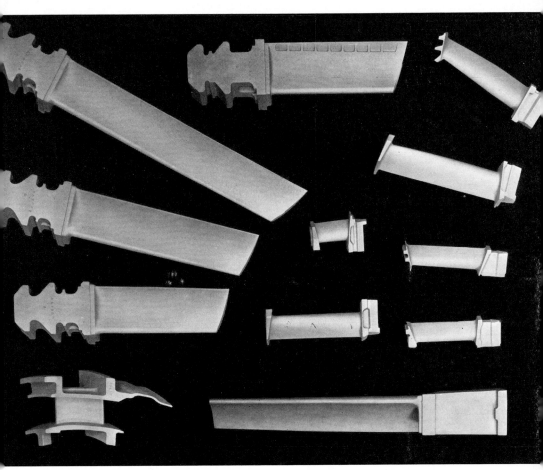

A group of gas-turbine blades and nozzle guide vanes, produced to high standards of accuracy in heat-resistant alloys that are very difficult to form. *AE Turbine Components Ltd*

Investment casting is generally used today for producing turbine components in the more exotic materials. Here the wax master for a cluster of vanes is being assembled prior to the moulding process. *AE Turbine Components Ltd*

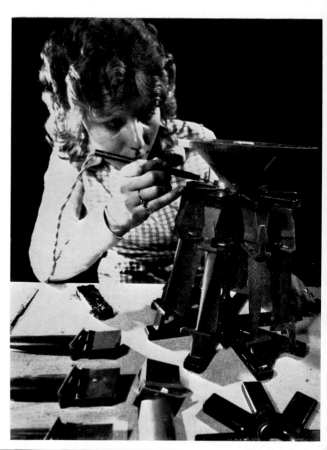

One of the latest joining techniques – electron-beam welding – being used in the production of the capsules that form pressure-sensing elements in the monitoring and control of gas turbines. *Smiths Industries Ltd, Aviation Division*

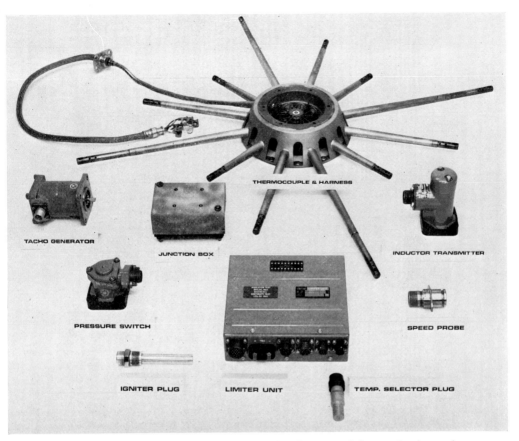

THERMOCOUPLE & HARNESS

TACHO GENERATOR

JUNCTION BOX

INDUCTOR TRANSMITTER

PRESSURE SWITCH

SPEED PROBE

IGNITER PLUG

LIMITER UNIT

TEMP. SELECTOR PLUG

The thermocouple spider and other components forming part of the monitoring and control system for the Pegasus gas-turbine engine of the Harrier 'jump jet'. *Smiths Industries Ltd, Aviation Division*

Semi-conductor material provides the path for the electrical discharge on this modern gas-turbine igniter plug. *Smiths Industries Ltd, Aviation Division*

Four-piece shell-type insert specially developed to facilitate the initial production and subsequent renewal of the epitrochoidal bores of Wankel engines. *Vandervell Products Ltd*

Automatic gauging of a gravity-diecast aluminium cylinder head for a Volvo car engine – part of the acceptance test procedure at the foundry. *Aeroplane & Motor Aluminium Castings Ltd*

A 6000-ton pressure-diecasting machine being used to produce the aluminium camshaft and tappet carrier for the Rover six-cylinder car engine. *Aeroplane & Motor Aluminium Castings Ltd*

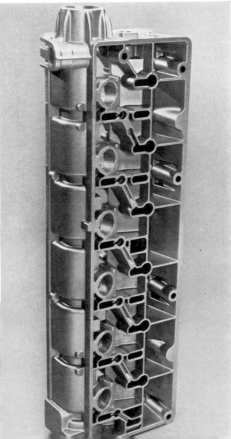

The finished camshaft/tappet-carrier casting; it is the largest pressure diecasting produced in the UK for the automotive industry. *Aeroplane & Motor Aluminium Castings Ltd*

Above: Final stages in the manufacture of a forged crankshaft for a locomotive diesel engine: the operator is lapping a crankpin to the finished dimensions. *GKN Forgings Ltd*

The Porsche 928 car engine has powder-forged connecting rods; these require minimal machining, as can be seen from this view of as-forged and finished-machined components. *GKN Forgings Ltd*

Opposite, above: A 400-ton press employed in the production of sintered components, many of which are incorporated in today's engines, especially in the automotive field. *Brico Engineering Ltd*

Opposite, below: A selection of components – including rockers, valve-seat inserts and piston rings – produced by the sintering process and requiring little or no subsequent machining. *Brico Engineering Ltd*

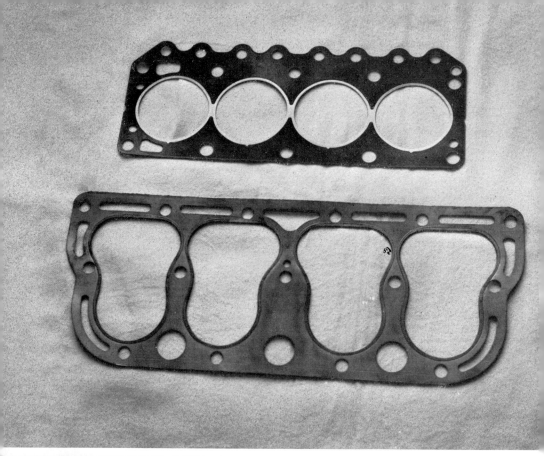

Cylinder-head gaskets of 1926
and 1978: the lower component
is a copper-asbestos gasket for a
'Bullnose' Morris engine and
the other, with steel bore rings
linked by a spine, is for a Ford
Fiesta unit. *Engineering
Components Ltd, photograph
by the author*

Paper elements are made in vast
numbers for filtering oil, fuel
and intake air; here, air-filter
elements are undergoing the
final stages of manufacture.
*AC Delco Division of General
Motors Ltd*

performance conditions. British ancillary-equipment manufacturers were again in the forefront here, and in 1945 Lucas introduced acceleration controls which ensured a constant maximum air/fuel ratio. One of these, a flow-type unit which varies the size of the fuel-metering orifice, is still used today in the USA, although in a rather more sophisticated guise.

Gas-turbine development was very rapid immediately after the war, and soon the engine makers – spearheaded as usual by Rolls-Royce – were demanding still more advanced control systems. In the early 1950s, therefore, all-speed governing of the fuel supply began to replace the previously used maximum-speed governing, and superior temperature-control arrangements appeared. The search for improvement went on, and Lucas engineers – in conjunction with the Rolls-Royce control specialists – decided that, since an engine is essentially non-dimensional in concept, the best form of control would be one that was based on non-dimensional parameters. They duly found suitable parameters (which, fortunately for me, are outside the scope of this book!) and designed a control system accordingly: it was called CASC – combined acceleration and speed control – and it was first fitted in 1958 to the Rolls-Royce Spey engine.

This type of control has been used, in one form or another, ever since, though Lucas added a pressure-ratio control for the Rolls-Royce RB211 turbofan engine in 1968 – the first time that any sort of 'thrust management' had been incorporated in the system. The additional control involved a pneumatic computer to resolve the complex pressure situation, and its necessity served to underline the fact that the limit of practicability was being reached for hydromechanical control systems, in which category come all those that have been mentioned so far.

Clearly, the future lay with electronic control, which does not imply that this was something new. On the contrary, electronics had been playing its part in gas-turbine control systems since about 1945 for such purposes as amplifying signals from speed and temperature measuring equipment.

Early amplifying units incorporated thermionic valves and so were relatively large and rather delicate. However, the advent of solid-state electronics in the early 1950s allowed much smaller and lighter designs to be produced, and the first transistor system was evaluated in 1954. There were still environmental difficulties, though, and the specialists had to learn a lot about encapsulating components in synthetic resins, and later in visco-elastic materials, in the search for long-term reliability.

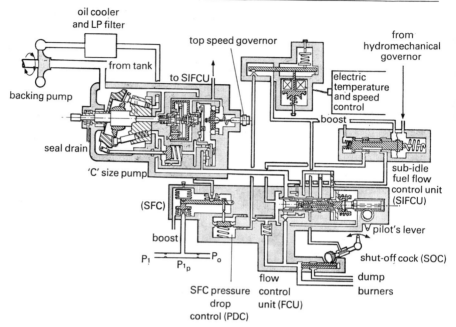

This schematic diagram of the Lucas Aerospace fuel system for the RB172 jet engine gives an idea of the complexities involved in feeding and controlling a modern aircraft gas turbine

During the last few years, considerable progress has been made in the development of full electronic control systems for aircraft gas turbines, and in Britain the protagonists have been Lucas Aerospace, Smiths Industries and Ultra. The trend is towards incorporating the electronic units on the engine, to make this a self-contained unit, although the environment is then clearly more hostile than in the case of remote mounting. Also, digital systems are becoming preferred to analogue, mainly because of the ease with which a digital control unit can be reprogrammed to suit a number of different engines. Lucas have been working on this aspect since the mid 1960s and their latest digital system has been switched among four gas turbines with 'softwear' changes only. At the time of writing, this particular design of control was undergoing flight tests on Concorde and was performing very satisfactorily.

Electronic control is especially suitable for helicopter duties owing to the need to govern rotor speed to very close limits; the complexity of the rotor dynamics is catered for better by electronics than by a hydromechanical control. Another major advantage of electronic systems is that they are the more compatible with 'full-authority' control which is coming into increasing use in all gas-turbine applications. In a full-authority system, the control unit operates automatically over the entire range of engine conditions, whereas with the earlier limited-authority type the operator is required to take over in certain circumstances.

In the introduction to this chapter, reference was made to the difficulties of controlling automotive gas turbines. Both here and in the industrial-engine field there has been some cross-fertilization from aircraft practice: a full-authority electronic control system, based on speed only, has recently been developed by Lucas for both these types of power unit as well as for light helicopters; it is known to have been undergoing evaluation by British Leyland on their truck turbine, and is to be fitted to a new US helicopter.

The point has already been made that control systems respond to engine speed and to certain temperatures and pressures, and it is usually desirable too for other temperatures and pressures to be monitored for information purposes. Consequently, a very important aspect of gas-turbine operation is the ability to measure all three parameters really accurately under all conditions.

In the measurement of rotational speed, only one fundamental advance has been made since the first aircraft gas turbines came on the scene. Initially the familiar tacho-generator – long used for the 'rev-counters' of piston engines of many types – served this purpose; in fact it is still quite widely fitted, mainly to American gas turbines. However, it has two basic disadvantages: the mechanical drive to the generator portion is not easy to arrange inside very high-temperature regions of the engine, while the drive at the output end can pose installational problems.

About ten years ago, therefore, Smiths Industries decided to develop an improved type of tachometer based on the inductive magnetic pick-up principle – known for many years but not previously applied to this activity. The pick-up unit of the magnetic speed probe, as it is called, lent itself admirably to 'burying' in a very hot environment, while the absence of any output drive simplified installation. There was one snag, though, in that the device did not produce sufficient power to drive a speed indicator directly, so a separate energy supply was necessary for this function. Nevertheless, the advantages were such that the magnetic speed probe soon became standard equipment on most British high-performance gas turbines.

For an exact knowledge of the temperature pattern through the engine, a number of readings have to be taken – from the ambient air entering the compressor to the gases in the jet-pipe or outlet. Following well-established industrial practice (rather than that evolved for piston engines, in which the critical temperatures are lower), thermocouple probes with nickel/chromium versus nickel/aluminium hot junctions have generally been used for thermometry in the hot areas. Few electrical changes have been necessary but substantial improvements have been made on the mechanical engineering side. New methods have had to be evolved of getting to the interior of the engine and of conveying the signal reliably back to where it is wanted. Another area in which the specialist firms have made a very significant contribution is in protecting the hot junction adequately against corrosion and vibration while ensuring quick response to variations in the temperature being measured. Any undue delay in recognizing a sudden rise in the temperature in the turbine rotor, for example, could have disastrous consequences.

The hottest parts of a gas-turbine engine are the inlet guide vanes to the actual turbine, because they are exposed to the gases as these emerge from the combustion chamber or chambers. On the larger engines the vanes are big enough for the temperature probes to be inserted directly into them or to form their leading edges. This is difficult if not impossible on smaller versions, though, so here the vane temperature has to be inferred from that in the jet-pipe which is more accessible.

It is usual to measure these high temperatures at several points over the cross-sectional area, because the temperature contours can be very complex; a single probe might give a very unrepresentative answer. Multi-point measurement means a harness to take the signals from the various probes via a common lead to the indicator or control box. Originally the harnesses of high-powered engines were veritable networks of cables, festooned around the engine and equipped with a number of electrical connections of sometimes dubious reliability. Once again, though, the accessory specialists stepped in to evolve something better.

The first move was the development of mineral-insulated rigid cables containing the conductors, the mineral being a magnesium oxide composition. Initial development work on mineral high-temperature electrical insulators was carried out by Smiths Industries at their Cheltenham and Putney factories, and the production technique they evolved around 1960 is still the best today. In it, the single or multiple conductor is inserted in a sheathing tube of a nickel alloy (Inconel or one of the Nimonics) and the annular space between is packed with the powdered mineral. The sheath is then swaged down in a machine to a considerably smaller diameter, thus compacting the powder into virtually a solid mass. After swaging, the sheath can be bent to any desired shape without risk of damage to the insulation.

This development made possible the robust 'integral' harness – a single assembly instead of a collection of bits joined together. Each harness of the new type had of course to be designed expressly for the individual engine, and here again substantial progress has been made during the past ten years or so. Smiths Industries' first integral harnesses for measuring jet-pipe temperature were mounted *externally*, with the probes projecting radially inward. Subsequently, though, to reduce the total length of conductors for the Rolls-Royce Pegasus engines of the Hawker Siddeley Harrier and the Olympus 593s of the Concorde, the harness was designed to be mounted *internally* in a housing forming part of the tail cone. The probes therefore project radially outward into the gas stream and, although the harness is exposed to higher temperatures than with the earlier scheme, it is less susceptible to damage in service.

Another aero-engine thermometry problem requiring solution was that of flame detection. In certain circumstances in the air, combustion can cease (in either the main combustion chambers or the reheat system) – with dangerous consequences unless the pilot is quickly informed of the situation and can initiate reignition. The difficulty is that, because of the considerable mass of metal in the turbine, the temperature falls relatively slowly after flame-out occurs, so an extremely sensitive detector is necessary. One system that has been introduced is capable of signalling flame-out within a tenth of a second and is impervious to extraneous messages. It is the ultra-violet detector which responds to the physical presence of flame in the combustion area, so the absence of response is the source of the signal.

The equipment used for measuring air, fuel and oil pressures, and for transmitting the signals, is no different in principle from that of motor vehicles. However, the temperature environment is much more severe (maybe ranging from as high as $+250°C$ down to $-60°C$) and robustness and a high integrity of the electrical connections are essential. Normally the pressure is applied to a capsule containing a bellows or diaphragm, the movement of which is employed to generate either an on-off or a continuous electrical signal, according to the duty.

Reliability demands careful design and construction of the capsules, and in [the early stages here one of the leading companies was Negretti & Zambra, the well-known makers of barometers and scientific instruments.

As a tailpiece on the control aspects of gas turbines, it is worth recording a difference that existed until recently between British and American aircraft gas-turbine engines. Because of the greater effectiveness and precision of their control systems, the British units could be operated closer to their limiting values of speed and temperature than could the US ones, which had to be derated somewhat to avoid the risk of damage. Realizing the performance penalty, however, the American manufacturers have made strenuous and successful efforts to catch up with British control standards and practice.

Ignition systems

Reference was made in Chapter 2 to the electric motors and other equipment used for starting aero-engines of all types. In the context of that generalized section, however, it was not appropriate to mention one aspect that is peculiar to gas turbines – the initial ignition of the fuel as it is sprayed into the pressurized air passing through the combustion chamber or chambers during run-up. This ignition requires a considerable amount of heat or other energy because of the fuel's relatively low volatility and large volume.

For the early turbines, such primitive devices as wax tapers and paraffin-soaked cotton waste – lit from outside the engine – were tried as igniters. These were obviously of limited practicability so the search began for something better. The first 'stabilized' ignition arrangement consisted of a large-gap high-tension plug, basically of automotive type, operating at a fast repetition rate by means of a vibrating-contact-type booster coil and positioned in the main fuel spray. It worked reasonably well but was not fully reliable because of the fouling that tended to occur once the ignition system was switched off when combustion had become self-sustaining. At much the same time there was some experimentation with heater plugs of the type used in indirect-injection diesels, but these too were not very successful.

The next stage in Britain, resulting from collaboration between KLG and the engine makers, was a 'torch igniter' in which an auxiliary jet of fuel was fired by a special sparking plug; the burning jet then ignited the main spray of fuel. Because of its thin jet of fuel, which was directed along the axis of the plug, the first torch system was not very effective, but it pointed the way for the development of an improved torch igniter that proved satisfactory in various engines and became a standard component for several years. However, it still suffered sometimes from the insulator fouling mentioned earlier, so alternative methods of ignition continued to be investigated.

In the late 1940s the torch igniter was joined by the first of the high-energy capacitor-discharge systems. This type of equipment, which did not require an auxiliary jet of fuel, proved superior for the more powerful engines then being built, and soon came into widespread use. Not only did it facilitate the initial starting of the engine but it enabled restarting after the dreaded flame-out that sometimes occurred in the cold air at high altitudes.

The high-energy systems were the result of collaboration between the electrical companies and the sparking-plug manufacturers; the former provided the source of the discharge and the triggering equipment, while the plug folk produced the actual igniters that converted the impulses into a succession of fat arcs (see later) in the engine. In the UK, Lucas and BTH played a leading role in developing the capacity boxes, as they are called, the igniter work being done primarily by KLG and Lodge. American ignition equipment and sparking-plug companies worked along generally similar lines.

Whereas the British systems featured a relatively low repetition rate – about 80 impulses a minute – right from the start, a substantially higher rate was adopted by Bendix in the USA when the Americans began to involve themselves seriously with gas turbines in the late 1940s and early 1950s. The two types still exist side by side today, neither showing any significant advantage over the other.

Capacity boxes are fairly straightforward examples of electrical engineering. They contain a bank of capacitors (condensers), a vibrator and a sealed gas-filled discharge tube which is triggered off at about 2000 volts, the current being taken from a battery. Stability of performance and long-term reliability are clearly essential features of these components which, in the case of aero-engines, also have to be as compact and light as possible.

The multiplicity of capacitors is necessary to obtain the high energy output necessary; this is usually 12 joules for a large aircraft engine, as compared with the 80–100 millijoules for the sparks in a vehicle engine. However, the voltage is very much lower than today's typical automotive equivalent of 20000 to 30000, since high voltages can lead to insulation breakdown – 'flash-over' – at high altitudes due to the low air density. This limitation is only an aircraft matter, naturally, but the HE systems were developed for aviation purposes and there was no point in subsequently evolving different designs when the established ones could certainly be satisfactory for all other gas-turbine duties also.

In comparison with the capacity boxes, the igniter plugs posed many problems. Basically they had to be of the surface-discharge type but more substantial than the automotive variety mentioned in Chapter 1 because of the much higher energy they had to handle. The first designs employed had an annular gap of 0·025–0·030 in. (0·635–0·762 mm) between the central electrode and the body, this gap being completely filled by mica insulation. However, the electrical discharge refused to jump across the mica, at the relatively low voltage, until fouling by combustion deposits formed a conducting path for it.

One way in which this reluctance was overcome was by filling the annular gap with a mica/metal-foil alternately wound laminate, to provide an intermittent path for the discharge. In an alternative development, the exposed end of a mica laminate was given a baked-on coating of colloidal graphite. The idea here, which was borne out in practice, was that, as the graphite was burnt away in service by the high-energy spark, it would be replaced by carbon deposits which would maintain a semi-conducting path; fuel/air mixtures were fairly rich at that time, so carbon deposition was correspondingly plentiful.

Such igniters provided an effective solution at the time and some engines are still using derivatives of them even today. In the early 1950s, though, combustion-chamber designs were evolved that gave improved overall performance, largely through the more even distribution of the fuel within the chamber. Consequently, the previously rich mixture in the vicinity of the igniter plug tip during normal running became considerably weakened, and this local weakening meant that carbon deposits no longer formed. The need therefore arose for surface-discharge materials that would retain their properties in the modified conditions.

KLG were the first to find a really satisfactory means of stabilizing the path for the discharge. This involved replacing the mica insulator by one made of the now-conventional alumina ceramic, and coating its end face with a fired-on semiconductive material resistant to high temperatures. By using this material it became practicable also to increase the spark gap to around 0·050 in (1·27 mm), with benefit to the relight performance. An added advantage was that any accumulation of combustion deposits did not impair the functioning of the igniter, as was the case with the earlier high-tension system; on the contrary, the deposits assisted the formation of the arc (rather than inhibiting it) by reducing the resistance between the electrodes.

Here, then, was the real long-term answer on ignition, so other makers soon followed suit, with variations as to the configuration of the central electrode and insulator, and to the material and method of application of the semiconductor; these variations included the use of a bonded-in pellet instead of a coating. It is interesting that the exact manner in which the semiconductor functions in this instance has still not been fully established. The generally accepted belief is that, when the electrical impulse is fed to the igniter, an initial 'leakage' of current takes place across the semiconductor surface. This current apparently heats the surface and ionizes the air adjacent to it, thus forming a conducting path for the main discharge which occurs as a loop-shaped arc lasting for about 80 microseconds.

Where this system is applied to land-based turbines, the energy level is generally lower because ignition at ground level is easier than at altitude where temperatures are very low. Typically 2, 4 or 6 joules are used, from boxes of basically similar design to those already described but usually adapted to be fed by mains electricity instead of batteries. This alteration necessitates a transformer and rectifier system as the capacitors must of course be fed with direct current.

As was stated earlier, high-energy ignition equipment does not necessarily demand a separate fuel jet. However, since today's requirements are for ever-greater combustion efficiency together with maximum certainty of relights at high altitude, the trend now is towards a combination of a high-energy igniter plug and an auxiliary fuel jet, the two forming an improved torch ignition system. This arrangement looks to be capable of meeting gas-turbine needs for some time to come.

Before we leave the subject, a few words should be said about 'reheat' (otherwise known as afterburning) – a means of increasing the thrust of a jet engine for short periods by spraying additional fuel into the jet-pipe. In the early days of reheat, that extra fuel had to be ignited and this was done with a second high-tension system since high-energy equipment had not yet become available. Nowadays, though, such an aid is generally unnecessary since the gas temperatures in the jet-pipe have become high enough for ignition of the fuel to be automatic.

The Wankel engine

Introduction

Despite its very short history (it was invented as recently as 1954) the Wankel rotary-piston engine is the only non-reciprocating alternative to the I C gas turbine to have reached commercial production. Regarded by some at one stage as the likely *successor* to the reciprocating I C engine, it has fallen somewhat from grace of late because of the disadvantages that accompany its considerable advantages. The Wankel now seems unlikely to oust its rival but will certainly coexist in some fields of application.

Ever since the early days of steam, engineers have sought to obviate troublesome inertia forces by substituting rotation for reciprocation in their power units. However, workable geometries were very elusive and, even if this fundamental difficulty could be overcome, the effective sealing of the rotating system remained a major problem.

Felix Wankel, an able German engineer, struggled with rotary I C engines for many years and refused to be discouraged by failure. Then in 1954 he rediscovered the feasibility of a triangular-shape rotor revolving within a two-lobe epitrochoidal bore of wide-necked figure-eight form; the housing revolved too but at a different speed, to give three chambers that alternately diminished and expanded in sequence. Here, clearly, was a practicable basis for a four-stroke engine.

The invention was taken up enthusiastically by N S U in Germany, with whom Wankel was collaborating at the time on the sealing of rotary valves – a matter of considerable relevance to the new project. During development work, induction and ignition difficulties were encountered because of the rotating housing, so N S U's Chief Engineer, Dr Walter Froede, evolved a highly ingenious 'kinematic inversion' in which the bore was stationary and the rotor orbited within it while turning on its own axis. Although the motion was no longer purely rotary, the practical gains were considerable, so this N S U–Wankel design became the definitive form of the engine.

N S U proved the principle on superchargers before embarking on an engine programme. Their first car with the new power unit went into production in 1964 and was followed three years later by the outstanding Ro80 saloon which had a larger, twin-rotor engine. By now a considerable number of well-known firms, being very impressed, had obtained licences to develop their own variations on the Wankel theme; among them were Toyo Kogyo (Mazda) in Japan, Curtis Wright and General Motors in the USA, Daimler-Benz and Fichtel & Sachs in Germany, Alfa Romeo in Italy and Rolls-Royce in Britain. There was much interest on the motorcycle side too, and Yamaha and Suzuki were among the two-wheeler licensees.

The Wankel's appeal lay in its smoothness, its compactness and low weight for a given performance, its few moving parts and its relative insensitivity to fuel quality. On the other hand it had both theoretical and practical disabilities, the latter revealing themselves along the thorny path of development. Its basic trochoidal geometry provided a long, thin combustion chamber of high surface/volume ratio, so the thermal efficiency was relatively poor; any attempt to improve this by raising the compression ratio only attenuated the combustion chamber further. On the practical side, too, there were ignition problems and the thermal gradient between the engine's hot and cold sides gave rise to distortional tendencies.

Hard work slowly overcame the practical snags and, when the USA's anti-pollution drive got under way, the Wankel's high fuel consumption looked an acceptable trade-off for its low nitrogen-oxide emissions – the result of its comparatively slow and cool combustion. The slow-burning characteristics also gave it a hot exhaust which, it was thought, could be used to help oxidation of the other pollutants – unburnt hydrocarbons and carbon monoxide. Then the US pressure on NO_x reduction was eased and fuel consumption came under the spotlight. The Wankel's poor showing here was largely responsible for an almost eleventh-hour decision by General Motors to shelve a very big project to build engines for certain of its cars.

Toyo Kogyo, who were producing Wankels at a much higher rate than anyone else, were less easily deterred, and their perseverance has led to substantial fuel-economy progress without sacrifice of the rotary's basic simplicity. Further consumption reductions could undoubtedly be achieved by increasing the complexity – by supercharging (perhaps exhaust turbocharging) or using a two-stage layout to increase the *expansion* ratio and thus raise the thermal efficiency; Rolls-Royce have already proved the practicability of the second expedient in a very advanced propulsive unit for tank duties.

The fact that US anti-pollution regulations (and those of some other countries) are to be tightened considerably in the early 1980s, could yet bring the Wankel back to its initial favour through the resulting low permissible levels of NO_x. In the meantime, though, the type can be expected to continue its advance in the 'leisure' field of motorcycles, powerboats, snowmobiles and so on, and in various low-power industrial applications where the virtues outweigh the disadvantages.

Though highly interesting as a new prime mover, the Wankel is almost a dead loss in the context of this book. Most of its components – carburettor or injection system, bearings and electrical equipment, for instance – are much the same as those used on reciprocating counterparts. The only major areas where the specialists' technology has really been stretched are those covered in the sole subsequent section of this chapter.

Housings, bores and seals

For reasons of low weight and good thermal conductivity, aluminium castings have been and are used for the housings of the big majority of Wankel engines, the remainder being of iron. Since these castings incorporate either cooling passages or fins (according to whether liquid or air cooling is employed), they are relatively complex and so have posed the founders some attendant problems. None of the latter have been sufficiently intractable, however, to necessitate the development of special techniques. No, the real difficulties in respect of these housings have lain in producing the correct epitrochoidal bore shape and rendering it adequately resistant to wear by the apex seals of the rotor.

Most of the credit for the progressive reduction in the wear rate of aluminium bores is due to the metallurgists and other materials technologists, but these experts have often been in the employ of the founders and the seal manufacturers (such as Goetze in Germany) who must therefore be given due praise. It is only fair to point out, though, that most of the bore-coating techniques and seals were not developed specifically for the Wankel. As so often, the secret of progress was hard work – the scientific evaluation of various combinations of existing materials and treatments to find the best compromise.

Having discovered that their earlier life-extending scheme of plating the working surface with hard chromium was not entirely satisfactory, NSU tried the Elnisil coating (the origins of which are not entirely clear although it was certainly worked on jointly by NSU and a German specialist company named Friedrich Blasberg) and developed it to a satisfactory level in the mid 1960s. This coating is in fact nickel containing dispersed particles of silicon carbide, a very hard material, and again it was applied by plating.

In contrast, Toyo Kogyo – the world's biggest makers of Wankel units, for some of their Mazda cars – adopted the 'transplant coat process' in 1967 for their first production engines. This process was evolved in the USA by Doehler-Jarvis, who are specialists in surface coating. In it, a thin layer of steel – 0·015 to 0·040 in. (0·38 to 1·0 mm) thick – is sprayed in molten form on to the core of the metal die used for producing the casting; the die is then preheated and the aluminium alloy poured into it in the usual way. The steel coating transfers itself from the core to the casting and is subsequently machined to its finished form before being given a thin plating of chromium for wear resistance.

However, in 1974 Toyo Kogyo went over to a different process, called the 'sheet-metal insert process' and developed by them, as a means of raising productivity. It involves a preformed steel-strip liner which is shaped around a mandrel and then its ends are joined; the exterior surface of the liner is roughened into a series of tooth-like projections. The liner is placed in the die, and the diecasting pressure causes a mechanical interlock between the projections and the aluminium alloy of the casing. After the casting process, the liner is machined and chromium plated as before.

In all these cases the epitrochoidal form has to be produced on special-purpose machine tools developed mainly in the USA. There are three basic types of machine – copy-grinding/cam-grinding, generating and contouring by numerical control. Although the German Kopp grinding machines used by NSU are relatively slow, they set the standard of accuracy by which others are judged.

However, all these machines are costly and at best the machining process takes some time to perform, neither of these factors making for a low-cost engine. Consequently, considerable importance must be attached to an ingenious application by Vandervell Products of the thin-wall bearing technique dealt with in some detail earlier, in connection with vehicle engines. Although the Vandervell development, announced at the beginning of 1974, has not yet been adopted by any manufacturer of Wankel engines, its potential for reducing production costs is such that it can hardly fail to establish itself before long.

The Vandervell system comprises a four-piece steel bore liner, the first advantage of which is that it fits into a space of 'arena' form in the casing. In this context an arena is merely two semicircles joined by two straight lines, and it is very simple to produce by the conventional and cheap boring and broaching methods. The four elements of the liner are two semicircular ends and two intermediate bridge-pieces. In normally proportioned Wankel engines, the ends of the epitrochoid are very nearly semicircles, so the true shape is produced in the liner ends by gradual variations in the thickness, for which Vandervell had already established the necessary production methods.

Thickness of the bridge-pieces varies also, but more substantially since they have to form the waist of the epitrochoid. They are positively located and attached by small bolts to the housing, whereas the ends are a push-in fit in the bore when there is the appropriate temperature differential between liner and housing, achieved by preheating the latter; the liners therefore have a slight interference fit in service. Initially Vandervell gave the rubbing surfaces adequate wear resistance by chromium plating, which presented no problems since it was done before the liner components were installed in the housing. However, the company has been investigating other and cheaper surface treatments, such as hardening by nitriding or its derivative Tufftriding, as a means of reducing costs still further.

To gain the maximum advantage from this invention, an engine manufacturer should really design his Wankel around the liner system. The necessary 'split' could then be in a plane through the minor axis of the epitrochoid – an arrangement that would have advantages over the orthodox one-piece main casting with separate ends. Not the least of these advantages would be the ease with which the elements of the liner system could be replaced in the event of their becoming excessively worn. Because of the accuracy with which they can be produced – as in the case of thin-wall bearings – there is no need for any finishing operation on the bore after installation of the liner.

Wear of the bore of a Wankel engine does not depend merely on the material subjected to rubbing by the rotor apex seals, any more than wear of the latter is a function only of the seal material. Apart from such influences as lubrication, rubbing speed and local temperatures, the overall wear picture is closely related to the compatibility of the bore and seal materials. Consequently, engine experimental departments investigating ways of achieving longer life have always been careful to change only one material at a time, and to assess the effects of that change thoroughly before trying anything else. It should be appreciated, though, that long life is more important for an integral bore than for sealings elements which are relatively cheap to replace.

The Wankel rotor has of course two types of seal – those at the tips, already mentioned, and those at the ends. Whereas the first-named are separators between the flanks of the rotor, the end-seals not only serve that purpose also but additionally prevent the loss of oil (from the rotor-cooling and/or bearing-lubrication systems) into the chambers. In spite of their single duty, the tip seals have the harder conditions since – owing to the geometry – their inclination to the surface of the bore varies continuously (as does the centrifugal force acting on them) while the rotor orbits and revolves. However, both types operate under the basic disadvantage of being single-stage seals, whereas the analogous piston rings of the conventional engine represent at least a three-stage seal. The maximum acceleration forces encountered by Wankel seals are lower than those on piston rings, although sliding velocities are higher.

Cast iron has always been the standard material for the end seals, but various metals and non-metals have been tried for apex seals. NSU first went into production with pressed hard-carbon seals, which are 'kind' to the bore surface, but they switched to a piston-ring cast iron in 1965. Their reason was that, for adequate strength, the carbon seals had to be relatively thick and so were too rigid to conform to a thermally distorted bore; as a result, wear was undesirably rapid and sealing less efficient than was required. Initially there were compatibility troubles between the iron seals and the bore, so the latter's coating was changed – as previously stated – from chromium plating to Elnisil. Subsequently, after numerous other experiments, NSU settled on sintered tip seals of a hard 'cermet' (ceramic/metal) called Ferrotic. It consists mainly of iron and titanium carbide and has proved to have a much greater wear resistance than its predecessor.

In respect of apex seals, Toyo Kogyo again showed a disinclination to follow blindly in the steps of NSU. They started with carbon seals and persevered with them much longer than did the German company, though they did change from a 'straight' carbon to an aluminium-impregnated version for which better self-lubricating and sealing qualities were claimed. To reduce manufacturing costs, however, they duly turned in 1973 to cast iron seals which were given good wear resistance by a chilling process that produced a very hard cementite layer on the crests of the seals.

We are unlikely yet to have reached finality regarding these small but critical components. The ultimate answer may lie in a 'space-age' material such as silicon nitride which, in the hot-pressed form developed in Britain by the Lucas group, is strong and tough enough to be used by

Rolls-Royce for gas-turbine blades. For such a duty it clearly has a high resistance to heat and corrosion, but in the Wankel context it has the added advantage of low frictional characteristics. Vandervell Products are among the specialist companies who have already done some exploratory work into the potential of silicon nitride, and they have considerable faith in it.

5 Areas of common interest

Introduction

In the preceding chapters, each component or accessory has been considered in relation to a specific type of engine, because of its clearly defined history of relevance to that type. However, certain other categories of 'bits and pieces' are incorporated right across the power-unit spectrum, being equally germane to all varieties. Since the importance of these categories in the overall picture could well be missed if their stories were tucked arbitrarily into one or another of the specialized chapters, I have given them what they deserve – a chapter to themselves.

Castings, forgings and sinterings

A very large proportion of the weight of any engine is made up of castings and forgings. Most commonly, the former are of iron, aluminium or magnesium (although steel castings are sometimes encountered) and the latter of steel or aluminium. Castings are made too – mainly for relatively small components – of zinc alloys, bronze and even brass; bronze and magnesium can be forged too, and the former is sometimes alloyed with aluminium for either casting or forging.

To avoid lengthy explanations. I must assume the reader to be familiar with the basic casting processes and with the elements of forging and its derivatives, stamping and extrusion. Sintering was explained briefly in Chapter 2 in connection with sparking-plug manufacture, and it is as suitable for a wide range of metals and their alloys as it is for ceramics.

It has the advantages of high dimensional accuracy and minimal wastage of material.

As implied in the opening sentence above, a vast range of engine parts is produced by these metal-forming processes. This is hardly the place for a detailed catalogue of applications, but some typical examples may be helpful. Engine castings include gas-turbine compressor housings (and sometimes the impellers and the turbine blades and vanes), cylinder blocks and heads, some automotive crankshafts, pistons and rings, camshafts, carburettor bodies and covers of various kinds. Forging and its variants are the usual methods for producing the more highly stressed components – heavy-duty crankshafts, connecting rods, valves, gudgeon pins, gears and some gas-turbine blades and vanes. Sintering, apart from established applications such as piston rings, bushes and thrust rings, is replacing stamping for various medium-duty parts – for example cams, valve seats, rocker pedestals, gears and sprockets for toothed-belt and chain drives.

By no means all engine manufacturers have their own foundry facilities, and then more commonly for ferrous than light-alloy castings, while very few have the capability for the other metal-forming techniques except machining. Consequently, large quantities of their castings and virtually all their forgings, stampings, extrusions and sinterings are produced by our old friends the specialists. Britain has long abounded in such companies – most of whom are situated in the West Midlands – and they rank very highly in the world league for know-how and quality; many therefore do a substantial export business as well as meeting home-market requirements.

Although sand casting is the oldest molten-metal technique, it changed remarkably little after the advent of the internal-combustion engine until the early post-war years. In fact, from the engine viewpoint, there were probably only two really significant steps forward between 1900 and 1945. One was the addition of linseed oil to the sand to increase the strength of cores, which are means of producing holes or hollow sections in a casting. This advance made it practicable to produce satisfactory water-jacket cores that could be removed after casting, thus considerably facilitating the manufacture of engine cylinder blocks. The other improvement was the mechanization of moulding and core-making, with considerable benefit to output rates.

As soon as the 1939–45 war ended, a real need arose, especially in the vehicle industry, to raise productivity and to save material weight and cost by achieving greater dimensional accuracy and a higher level of process control. The iron and aluminium founders therefore got down to a serious study of ways and means. One of the first improvements to be adopted was 'shell moulding'. It was the result of investigations that were under way about 1939, in Britain, the USA and Germany, towards finding a casting method of high accuracy but simpler and more economical than investment casting. The breakthrough was actually made in Hamburg by Johannes Croning who first hit on the idea of using a thermosetting synthetic-resin powder to bond the moulding sand into an integrated mass.

Shell moulding is basically very simple. A metal pattern of the object to be cast is heated, and sand with the appropriate proportion of the powdered resin is poured on to its surface. (Initially an intimate mixture of sand and powdered resin was used, but later practice has been to pre-coat the individual grains with the resin, this being done in a separate plant.) The heat causes the resin to flow and bond the sand grains together, so a thin partially cured shell of sand/resin is formed on the pattern. When this shell is thick enough, the pattern is inverted and the surplus sand falls off for re-use. Pattern and shell are then transferred to an oven where curing of the resin is completed; next the shell is stripped off the pattern which is returned for the preparation of the next mould.

Some castings require two or more shells, and these are secured together by clamps, bolts, tapes or glue. They then have an indefinite storage life, as long as they are carefully treated, so can be produced in batches against future requirements. It may be necessary to reinforce the shells against the loadings produced when the metal is poured; this is done by embedding them in sand or metal shot in a box. The permeability of the shells allows air and gas to escape readily, which helps towards a good metallurgical structure. Finally the shells are broken away to release the casting.

Shell moulding is now operated by many founders and is suitable for iron, aluminium, bronze and zinc. It is not the answer to all casting problems, but it does lend itself well to the production of a number of engine components, notably crankshafts (usually in spheroidal-graphite, or SG, iron), cylinder liners and air-cooled cylinder barrels and heads, where its precision enables thin and closely pitched finning to be employed. Because of the in-filling effect of the resin, too, the surface finish of a shell-moulded casting is much superior to that of an ordinary sand casting.

The shell-moulding technique has been adapted also to the production of cores for the inlet and exhaust ports in cylinder heads, the cylinder bores and the water jackets in both heads and blocks. Core-making methods have advanced enormously, and the top-grade foundries can now produce one-piece jacket cores, and port cores, for some small and medium-size piston engines, and have considerably reduced the number in many other instances.

A recent development here has been the making of cores in a somewhat similar fashion but without the need of heat: the sand is mixed with a resin and a hardener immediately before use, the hardener causing the resin to cure cold. In an alternative cold-cure process, developed by the Ashland Chemical Co., the sand and resin can be premixed at any time because the curing is effected after their delivery, by passing a volatile amine through the mixture in the corebox. The next general advance in forming the porting in aluminium cylinder heads for car engines could well be the use of steel cores instead of sand ones, for still greater precision. This foundry technique is already used in the production of heads for one leading European company which is apparently well satisfied with the results.

In parallel with these advanced methods of producing cores have gone increasing automation and locational and other improvements in casting techniques. British companies – Birmid and others on the iron side, Aeroplane & Motor Aluminium Castings and others on aluminium – have been in the forefront of this progress, and the result has been lighter, cheaper and better castings. A bit of a fuss was made a few years ago about the US development of 'thin-wall iron castings' for vehicle-engine cylinder blocks, as a better proposition than aluminium castings for reducing engine weight. In fact, though, the Americans were only then catching up with what the British (and other European countries) had already been doing for years!

The gravity diecasting of aluminium was introduced between the wars because of the deficiencies of contemporary sand-casting techniques in respect of speed of production and the accuracy and mechanical properties of the castings themselves. Many companies adopted the process, including AMAC (previously mentioned), High Duty Alloys, Birmingham Aluminium and Sterling Metals in Britain, and in general, they achieved their objectives. However, the subsequent progress in sand-casting methods has reduced the advantages of gravity diecasting except in respect of the product's mechanical properties.

Pressure diecasting, on the other hand, is finding increasing favour although it has a rather shorter history. It is employed for producing aluminium or zinc-alloy components and was originally adopted for the really large-volume manufacture of relatively small but complex and precise components – carburettor bodies, for example. However, since about 1955 its usage has climbed well up the size scale, to embrace quite large items such as aluminium cylinder-block/crankcase castings for car engines. This extension of the applicational spectrum is due largely to the excellent mechanical properties and freedom from porosity that result from the high pressures exerted on the molten metal in the die cavity. Castings of such a size would not have been practicable without the improvements that have been made in three respects – the manufacture of the dies themselves (by milling and other methods), the understanding of casting requirements in respect of optimum metal flow and cooling rates, for instance, and the consequent ability to 'design for the process'.

Since investment casting was covered in some detail in Chapter 4, there is no need to say any more here about the actual process. It is worth recording, though, that here again the technique has been adapted for the production of larger components than were thought practicable a few years ago; the complete impellers for the radial-flow compressors of small gas turbines are typical of what can be achieved by the experts today.

The traditional hammer-forging and die-forging processes continue in widespread use, albeit with many detail improvements through the years. However, they have been joined during the post-war era by more advanced methods that have been pioneered mainly by the big companies such as GKN Forgings, often in conjunction with research establishments or university laboratories. These modern techniques come under the general heading of 'precision forging', and their objective is to reduce wastage of material, both at the actual forging stage and in machining afterwards.

One such process, introduced in the late 1960s, is impact machining – a hot-working technique involving two stages, in the second of which the item is given its final form between very accurate dies. Another, with a rather longer history, is cold extrusion in very powerful presses; this extrusion can be either forward, through an orifice-type die, or backward, where a tool is forced into the billet (which is within a female die) and the metal flows the other way into the annular space between die and mandrel. This technique is used in the Associated Engineering group for the production of gudgeon pins.

The most recent additions to forging practice are the high-energy-rate processes. Earlier HER machines, operated by compressed gas, were very large and were introduced for the rapid forming of strong and 'difficult' materials – such as nickel alloys and titanium – for gas-turbine and other aerospace applications. During the last few years, though, smaller machines with a wider range of applications have been evolved. One of these is the Petroforge which is actuated by the ignition of a petroleum-based fuel and was invented in the engineering laboratories of the University of Birmingham. At the time of writing it was undergoing commercial development.

Unfortunately, the earlier promise of high-energy-rate forging has not yet been borne out in practice, one of the reasons being the short life of the dies on account of their adverse operating conditions. Efforts to use the process in the aerospace industry for the production of large gas-turbine blades and discs have been disappointing also, for a different reason: the exotic materials used for these components proved sensitive to the rate at which they were deformed, a very high rate tending to cause cracking. The most satisfactory method here is therefore still hammer forging which results in step-by-step deformation.

In the field of powder metallurgy, also, many improvements have been made, enabling a more extensive range of metal parts to be produced economically by the sintering process. The production engineers must be given credit for some of this progress because of their development of ingenious tooling and powder-compacting presses which have removed some of the earlier constraints of shape and complexity. On the metallurgical side, though, considerable advances have been made on four fronts – in the development of sintered steels of greater strength and ductility (without any loss of the dimensional precision for which the process is famous), in the selection of the alloy system to be used, in improving sintering techniques and in reducing the residual porosity.

Minimal porosity has been found to be essential if the toughness of unsintered steels is to be matched, and it is achieved by the use of super-compressible metal powders and high compacting pressures, and by re-pressing after sintering. In fact zero porosity can be obtained by re-pressing at an elevated temperature – a technique that has been variously called sinter forging, powder forging or hot recompaction. This technique is considered the 'ultimate process route', since its use enables the mechanical properties and accuracy of precision forging to be combined with the cost savings associated with the powder-metallurgy production of complex shapes. It is noteworthy that automotive connecting rods are now produced in small quantities by powder forging as an alter-native to conventional forging, and this application (of which the G K N group is one of the pioneers) is likely to increase in the future.

The closed-die tooling normally employed for these various methods demands that the high compressive forces are applied in one direction only – that is, uni-axial compaction. Consequently there is a limit to the variety of shapes that can be produced. One way out of this restric-tion is to use isostatic pressing, where hydraulic pressure is applied in all directions to the metal powder contained in a rubber mould. In addition, high-energy-rate compacting techniques – explosive or magnetic – are being explored as alternative means of reducing the residual porosity or of compacting special alloy powders which have poor pressing characteristics. However, at present H E R compaction does not seem to offer any special advantages in the field of engine components.

The value of collaboration between the engine designer and the specialist supplier has been pointed out several times already in this book. However, collaboration is so important in relation to castings, forgings and sinterings that I make no apology for raising the matter again. So often in the past an engine company has gone to the foundry or forge and said, 'This is what we want – make it!' and has there-by committed itself to unnecessarily high costs, because of consequent production problems, and maybe to an inferior article also.

An earlier approach, on the lines of, 'This is what we would like – how does it grab you?' would bring all the casting/forging experience to bear. A change of draft-angle here, an increase of radius there, a modification to the shape of a boss, a drilled passage rather than a cast-in one, even the use of a different material specification – these are the sort of alterations that the specialist will suggest to ease his own task and so ensure a better component at a lower price.

Gaskets and sealing washers

Joint-sealing is one of the least spectacular but most essential areas of engine technology. Whatever the type of power unit, it has numerous bolted-up joints which have to contain water, fuel, oil or high-pressure gases, and leakage past the sealing medium can result in a variety of troubles ranging from a messy exterior to complete failure of the engine. Although quite a lot had been learned about sealing methods on steam plant before internal combustion appeared on the scene, the IC engine brought with it a number of problems not previously encountered. Once again, the specialist companies had to find the solutions – not just initially but on many subsequent occasions as power outputs have soared and operational conditions have become more arduous.

To most people the term 'gasket' means 'cylinder-head gasket', and certainly this component has one of the toughest environments of any type of seal on a vehicle engine. Since earlier designs featured integral cylinders and heads, there was no need for such gaskets until the advent of the Model T Ford in 1908, after which time the detachable cylinder head quickly became the rule rather than the exception. From its birth, though, the head gasket was based on compressed asbestos fibre because of its heat resistance and compressive strength.

In the early 1920s one of the jointing manufacturers, believed to have been Payen or Cooper's Mechanical Joints, hit on the idea of strengthening these gaskets and improving their thermal conductivity by sandwiching the asbestos between two thin sheets of copper; the holes for the cylinder bores, and the transfer passages for water and oil, were reinforced with copper or brass eyelets secured under a press. Copper–asbestos gaskets were soon generally adopted and continued virtually unchanged until after the Second World War. The initial post-war advance by the gasket makers was to reduce the thickness drastically, in some cases by as much as 50 per cent to around 0·030 in. (0·76 mm). Such reductions cut the cost a bit, as did a design improvement that deleted the separate eyelets: one of the two copper sheets had plain edges to the holes and the other had flanged edges; during the sandwiching operation the flanges passed through the plain holes and were then lapped over. One of the first British firms to make gaskets of this kind was Payen, later to merge with its former competitor, Cooper's Mechanical Joints and eventually to come under the wing of the big Turner & Newall industrial group.

Developments of a more revolutionary nature began to take place in the 1950s, however. Copper was becoming undesirably expensive for most gasket applications, while many engine producers were endeavouring to save weight (and cost) by cutting down the wall thickness of their castings. These skinnier castings were also usually less rigid and hence more liable to distortion, thus posing greater problems for the gasketeers. The first of the resulting 'new generation' of gaskets appeared in the USA, where the casting rigidity difficulties were greatest. It was of tinplate (tin-coated sheet steel) about 0·015 in. (0·38 mm) thick and of corrugated form; to obtain a seal around the bores and water passages, the lips of the holes were given a shallow arcuate section.

Since that time, fewer and fewer new gasoline and diesel automotive engines have incorporated copper-faced head gaskets, although considerable numbers of these are still made as replacements for units of earlier design. The corrugated steel gasket was only one of the many different types that have been developed over the past twenty years. Contrary to some people's expectation when that steel component first appeared, compressed asbestos fibre is still very much to the fore, often combined with tinplate in a wide variety of ways. In some instances the CAF now incorporates a proportion of synthetic rubber to improve conformity to the two surfaces without loss of dimensional stability.

Some present-day head gaskets for vehicle engines are basically of CAF, with tinplate eyelets – often of the 'backbone' or continuous type, instead of individual, for the bores. Others are faced (sometimes on one side only) with tinplate, which frequently is lacquered to improve the sealing characteristics, or embody it as a sandwich layer. All these layouts are common to both gasoline and diesel automotive engines, but the latter category includes further varieties including laminated and solid steel gaskets and individual steel rings at the bores only. As already pointed out, diesels tend to produce higher combustion temperatures and pressures than do spark-ignition ones, so for strength their gaskets usually involve a higher proportion of metal.

In the aircraft field, the practice has long been to use individual bore-sealing rings, and these are usually still of copper–asbestos because price is less critical than for vehicle-engine components. Racing-car engines (another 'special case' in cost terms) have for some years also had individual fire-rings, as they are called, between head and cylinder liners. In this instance, though, they are of laminated steel instead of copper–asbestos. Rings are commonly used too for the big diesels in the industrial, marine and locomotive categories. Steel–CAF is the most popular combination, and an unusual configuration – evolved for seals in the petrochemical industry – has recently been adopted for highly turbocharged engines: the rings consist of alternate steel and asbestos strips, spirally (not helically) wound so that they are subjected to edge loading.

It will be apparent from the foregoing that the whole approach to cylinder-head gasketry today is both broader-based and more scientific than of old, and the engine companies' confidence in the specialists is measured by their rarely feeling the need for consultation during the early stages of designing a new power unit. They expect the gasket firms to be able to cope, as in general they have, with any set of basic require- ments plus development changes such as increased bore diameter (giving reduced space between the bores) and higher loadings through uprating.

Modern equipment has greatly helped the suppliers of these compo- nents to meet ever tougher demands. For example, they can now obtain accurate pressure patterns of clamped-up head/block combinations, enabling them to ensure that sealing is effected in low-pressure areas, while the adoption over the past few years of automatic test program- ming for engine dynamometers has brought a high degree of operational realism into the laboratory and thus eliminated many thousands of miles of development running on the road.

As indicated initially, there are many other types of joints in engines, and here also the specialist companies have taken advantage of the widening range of suitable materials. CAF is found of course in most high-temperature applications, but in addition we have paper (with or without synthetic-resin impregnation), rubber and cork, used singly or in various combinations. Much of the rubber employed today is of the synthetic type, since natural rubber has little resistance to mineral lubricating oil. A particularly successful post-war combination has been granulated cork bonded with synthetic rubber; gaskets of this material have replaced the earlier cork ones – with a fabric backing or inser- tion – for sealing rocker-box covers, oil sumps, etc., and they are thinner and have a much longer effective life than their predecessors.

Filtration

The need for adequate filtration of the 'vital fluids' – air, oil and fuel – is common to virtually all varieties of internal combustion engine, although the criticality of the various aspects depends to a considerable extent on the power-unit type and the operating conditions. For example, fuel filtration has to be more meticulous in a diesel engine – primarily because of the possibility of injection pump wear or damage – than in most gasoline engines. Again, more effective filtration of the air is required in dusty desert operations than in a temperate climate where any dust is frequently laid by rain.

In the days before the Second World War, filtration of all three types was relatively primitive. If an air filter was fitted at all it was rarely more than a screen of wire gauze, the purpose of which was as much to exclude relatively large objects, such as nuts, bolts and the end of a mechanic's tie, as to prevent the ingestion of abrasive particles from the atmosphere.

Oil filters, too, generally comprised no more than a simple element of wire gauze or perforated metal, situated either in the filler orifice or, in the case of dry-sump engines (of motorcycles for instance), in the base of the tank at the take-off point. However, centrifugal oil filters – incorporated in the crankshaft – did come into use for aircraft engines in the 1920s and they worked quite well because of the considerable difference between the densities of grit and lubricating oil. Then in the mid 1930s the automotive filter makers became a little more scientific and introduced external oil filters containing flannel-bag or mineral-wool elements. Because of the high flow-resistance of these, though, they had to be inserted in a bypass circuit – not in the main supply line – and only about 10 per cent of the oil was actually going through the filter at any one time; filtration efficiency therefore suffered.

Much the same situation existed in respect of filters for diesel fuel, in that felt or cloth elements had come into general use to keep abrasive particles out of the injection pump. In this duty, though, it is clearly necessary to filter *all* the fuel on its way to the engine, so relatively large elements were required to obtain the desired rate of flow.

The Second World War showed up the inadequacies in engine filtration, as in so many other aspects of technology. Many vehicles and aircraft had to operate continuously in particularly arduous desert conditions and with low standards of maintenance, yet they had to be both reliable and durable. Because of the pervasiveness of the fine sand, wear rates in the earlier stages of the desert war were frighteningly high, as was the consequent unserviceability, so the specialist firms had to take immediate and rapid action.

Although considerable progress had to be made during the war years, it was in respect of techniques rather than media. As an example, the experts developed two-stage air filtration which became widely employed on desert vehicles, both fighting and back-up. The first stage was a centrifugal or 'cyclone' pre-cleaner analogous to the oil-filtering arrangement already referred to, and the air from this was passed through an oilbath filter. In the cyclone cleaner, the air enters tangentially at the periphery of the casing and so is given a swirling motion which causes the heavy solid particles to be centrifuged to the outside where they are collected.

The oilbath filter was actually a development of an earlier type in which the air passed through an oil-wetted steel-gauze element of the 'pot scourer' type; since the element was pre-wetted it was effective only for the short period until the oil dried out. However, by mounting the gauze element above an oilbath, so that it was continuously wetted and washed free of dirt, more efficient filtration was obtained over a much longer period – until the oil became so charged with dirt that it had to be drained and renewed. It is noteworthy that this two-stage principle is still utilized today on some tractors and earth-movers, particularly in countries where the available labour is of a low standard.

Modern filtration, in all three categories, really began in about 1950 with the advent of the paper element. Credit for the original invention appears to lie with the paper companies, but that for its application is shared in Britain by a number of filter manufacturers including CAV, Coopers, Vokes and General Motors' AC-Delco subsidiary.

Paper is of course a porous material, and it proved possible – after several years' work – to endow it with very tightly limited and consistently maintained particle-trapping capabilities. However, it is also an inherently flabby material, so a stiffening agent was necessary; right from the start, impregnation with phenolic synthetic resin proved a completely satisfactory means of reinforcement. Although paper was a more effective filtration medium than felt or cloth, it was found to have a greater propensity for clogging. To ensure an acceptable service life of the element, the latter had therefore to have a larger area than had previously sufficed, so the immediate problem was how to accommodate that area effectively in a compact container. The difficulty was particularly acute in the case of automotive oil filtration: full-flow filtering of the lubricant was by then becoming recognized as much superior to the bypass layout, so a big increase in area was necessary on that account too (see later).

Since much of the paper-element technology was consequently developed in connection with oil filtration, it is appropriate to look initially at this aspect. First came the bellows or concertina type of paper element, usually of octagonal form. Since double folds were unavoidable at its corners, however, these were heavily stressed and failures therefore occurred of the adhesives used to bond together the constituent paper folds. In addition, the folding was not only troublesome in production terms but also prevented access of the oil to parts of the element, so this had to be appropriately enlarged to provide the necessary effective area.

These disadvantages were overcome by the multi-disc type of element, introduced in about 1958. It featured interleaved paper discs which were embossed to maintain clearance between one and another; in effect, each pair of discs formed a hollow fin, edge sealing being effected by the central axial loading on the pack. Oil was delivered to the container on the outside of the element, passing through this to the inside. The disc-type filter performed well provided the central holes of the disc were really accurately positioned, to ensure alignment, and it was appreciably smaller *pro rata* than the bellows design. It lent itself to automated production in large quantities, but a lot of paper was wasted in the blanking of the discs.

Production economy was therefore the principal reason behind the next step; this, taken in the 1960s, was to the vertically pleated element which has become virtually the standard type for oil filters. Such an element is produced from a single strip of paper with a glued joint along a pleat; the paper is pre-creased for easy and accurate pleating, and is embossed to ensure the correct interpleat spacing on both the inlet (dirty oil) and outlet (clean oil) sides. After pleating, the element is formed into the required shape – normally cylindrical for oil filters – and is retained in that form by end pieces.

This particular kind of pleated element is eminently suitable for highly mechanized and automated production techniques, and it is quite compact in relation to its effective area and flow rate. Where space is at a particular premium, though, some makers use a special W-fold which can reduce the occupied volume by up to 30 per cent.

Until relatively recently, oil filters with paper elements were of the 'detachable' type – that is, the element was a separate component from the housing which was a permanent part of the engine. At oil-change times, the housing was detached and the element renewed. Since the early 1970s, however, the 'spin-on' cartridge filter has become almost universal for car engines and is beginning to penetrate the commercial-vehicle market – possibly as a prelude to its adoption on still larger engines although it has less advantage in their case.

The cartridge filter comprises an integrated element and canister which are thrown away complete at oil-change time. This may seem an uneconomic arrangement in respect of both manufacturing and servicing costs. In practice, though, it works out well enough, the slightly higher price of the integrated unit being offset by the quicker replacement that results.

Because these modern filters are of the full-flow type, a relief valve has to be incorporated in the system to obviate any risk of failure of the oil supply should the filter be kept in service for so long that serious blocking occurs. Today that relief valve is sometimes embodied in the filter unit. The specialist manufacturers produce a range of filter sizes to cater for different flow rates, but for production economy they encourage the engine builders to standardize on only a few sizes, even if this results in a smaller power unit being 'over-filtered'; multiple filters can of course be employed for larger engines if necessary.

In view of this present-day sophistication, it should be recorded that the centrifugal oil filter, mentioned early in this section, is still flourishing after more than fifty years. That a range of such filters should be made by Glacier is particularly appropriate since dirt in the oil remains the primary cause of bearing damage. By removing solid particles down to a mere 2 microns, the Glacier units are making a useful contribution to bearing longevity through harnessing centrifugal forces as high as 1500 times the gravitational force. At the bottom of the size scale are filters for car-type engines, while at the top is one for systems circulating about 300 gallons (1400 litres approximately) of oil: this big unit has a capacity for no less than 24 lb (11 kg) of solids.

A noteworthy final point on oil filters is that, by a coincidence, the need to fit higher-capacity filters to cope with full-flow systems was offset to a considerable extent by the introduction of 'dispersant' oils at the end of the 1950s. As a result of the dispersant additive, fine foreign particles – such as carbon – remained harmlessly in suspension in the oil instead of coagulating into the larger masses which previously had been collected by the filter. This simple fact had a dramatic effect on the filtration area needed for a given duty, the decrease being over 50 per cent. When reducing filter size accordingly, though, engine manufacturers had also to stipulate the use of dispersant oils to avoid the risk of premature clogging of the element.

As previously indicated, fuel filtration for most gasoline engines is not a critical operation. The majority of such power units have their mixture delivered by a carburettor or carburettors; these instruments are not liable to significant wear or damage through solid particles in the fuel, so the system merely has to trap those big enough to block the smallest jets. Consequently there is rarely the need for anything more than a bowl-type filter (which traps by gravity the larger particles and any water in the fuel) and maybe a small fine-gauze thimble at the float chamber. However, AC-Delco and others have recently developed very small paper-element petrol filters for in-line installation.

While the needs of carburettors and some petrol injection systems are met by these small in-line filters, other PI designs are as sensitive to foreign matter as is diesel equipment and they require similarly comprehensive filtration. Injection-pump wear in either case can mean very difficult starting as well as a fall-off in performance, and so has to be kept to a minimum. To underline the inefficiency of diesel-fuel filtration when the Second World War began, incidentally, pump wear in the Western Desert was so acute at first that the British High Command seriously considered abandoning the diesel engine for North African duties!

The palliatives were found, as already indicated, and research began into the next generation of diesel filters – based of course on the paper element. Much credit must go here to CAV whose research team – headed by Dr A. E. W. Austen – established the criteria of pump wear scientifically in the late 1940s and led the way in developing the new filters. Although a good filter medium, CAV's paper had a typically low transmission and so needed a large area to avoid the big pressure drop that would cause malfunctioning of the injection system or necessitate short replacement intervals. The company evaluated a considerable number of element configurations in the search for compactness, and finally evolved its own coiled design which was patented and has been used consistently since 1950. This coiled element is not suitable for oil filtration, because it could not withstand the relatively high pressures in lubrication systems, but it is excellent for diesel fuel since its ratio of area to occupied volume is nearly 50 per cent better than that of the vertically pleated variety described earlier.

The CAV element has always been sealed into its canister, but originally the flow was upward through it, leaving a sedimentation space at the bottom. However, this layout resulted in a tendency for entrained water to be taken through the element, going on into the injection equipment in sizable globules. Because the consequent corrosion was undesirable, the company switched to what is called 'agglomerator flow' – down through the coil and up again centrally – thus allowing the water to collect in the bottom of the canister.

Most manufacturers of diesel fuel-system equipment have evolved supplementary units to enhance the performance of the main filter. These units, which come under the general heading of sedimenters, are in effect more complex derivatives of the bowl-type petrol filter already discussed, in that they work gravitationally, without imposing any resistance on the fuel flow, to remove large solid particles and/or water.

They have become widely used where fuel contamination is prevalent, either in bulk storage or in transfer to the engine's tank.

The most recent advance in this area, again from C A V, has been the development of devices that monitor the water collected by a sedimenter and give a warning when it exceeds a predetermined level, so that it can be drained off before there is any risk of water being drawn into the engine. Electronic and mechanical layouts have been evolved, for both automotive and industrial applications.

In air filtration, the paper-element type began to replace the dry gauze, oil-wetted and oilbath varieties in the mid-1950s. Again, the filter manufacturers have often had to cope with a small space availability – a situation exacerbated in the vehicle field by the recent trend towards lower bonnets on cars, and often also by the tendency of their designers to ignore the air filter until almost the last stages; the filter company then has to produce an awkward (and usually unnecessarily costly) shape of body to fit into the only space left!

The pleated element has become general for air filters, on both gasoline and diesel engines. Elements for the former are usually of cylindrical form, or somewhere near (certainly for carburettor installations where the filter normally sits horizontally on top); the air is

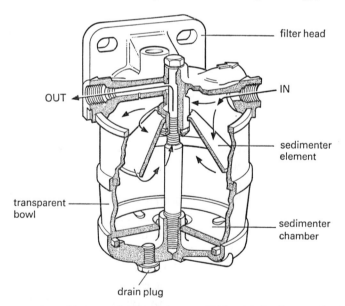

Sedimenters such as this Lucas C A V unit are specified on many diesel engines to remove water and larger particles of foreign matter from the fuel before the main paper-element filtration

taken in at the periphery and delivered centrally. Similar units are employed too on most automotive diesel engines, but for larger diesels panel-type filters are often preferred. They have a flat element or elements and may incorporate a flame-trap and perhaps a fuel-spray system to facilitate cold starting.

An interesting refinement, pioneered by AC-Delco, is to coat the element with a non-drying fluid to enhance the dust-retention capacity corresponding to a given increase in the pressure-drop across the filter. The fluid coating causes the particle layer to build up on the surface of the paper instead of becoming embedded in its structure, thus delaying blocking of the element. Parallel dust tests have shown that an element so treated can retain about three times as much dust, by weight, as an untreated one for the same pressure-drop change.

This treatment is beneficial too in considerably reducing the blocking effect of exhaust carbon which is always prevalent on the road and increases with the traffic density. Some present-day underbonnet conditions, as already indicated, impose such severe restrictions on air-filter size – and therefore element area – that without such treatment the blockage time could be embarrassingly short. Treating elements with a non-volatile liquid can also impart flame-retardant properties, and Coopers (whose work on this aspect paralleled that of AC-Delco on life extension) have made extensive use of the fact for car filters since the late 1950s.

The earlier reference to two-stage air filtration requires some augmentation in view of the contribution made by the paper element. A disadvantage of the cyclone/oilbath layout is that its efficiency varies with the throttle opening which modifies the airflow through the filter. Also, the separating ability of the oilbath section diminishes as the oil thickens with collected dust. Both deficiencies were overcome in the late 1950s by replacing the oilbath by a paper-element second stage.

Most sophisticated of these improved filters were those developed by Coopers; they had multi-cyclone first stages and long-life cleanable paper-element second stages, and soon began to find favour for heavy-duty off-highway vehicles. Similar assemblies but with less elaborate first stages, were adopted for on-highway and medium-duty off-highway operations. Coopers were also responsible for two associated engine-protection devices: one was a visual restriction indicator, which enabled the driver to tell at a glance when the element required servicing, and the other was a safety element that was automatically brought into action if the main element was damaged.

A variant of the Coopers multi-cyclone system was developed specially for tracked military vehicles. It had a two-stage paper-element section in tandem with the cyclones, from which the centrifuged dust was scavenged by a patented system utilizing an air-bleed from the pressure side of the engine's turbocharger. The same sort of practical thinking has gone into the company's latest filter for heavy vehicles; it has a cloth element which is automatically cleaned as necessary by the reverse flow of compressed air from the engine-driven compressor.

Reference was made in the section on vehicle carburettors to the recent introduction of 'environmental' devices to provide thermostatic control of the inlet-air temperature. Some of these devices fall very much into the ambit of the filter makers. In certain instances the latter embody the engine firm's ideas in the components they produce, but A C-Delco have patented several arrangements, one of which is a manually set summer/winter device with an automatic over-ride which, in the winter position, bleeds in some cold air at higher engine demands.

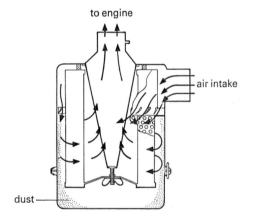

Two-stage air-filter arrangement evolved by Coopers for automotive-type engines operating in moderately dusty conditions

Another current trend, the motivation for which has come from the filter manufacturers, has been towards the use of plastics instead of steel for air-filter bodies, particularly for cars. The principal objective here has been to cut production costs, in that a one-piece moulding replaces an assembly of several pressings. However, good design can also effect a reduction in under-bonnet noise since plastics are less 'live' than steel in acoustic terms. Some of the filter companies produce their own mouldings (as well as their own steel pressings for other designs), and one at least has become so expert as to supply the relevant tooling to others less well endowed.

Epilogue

I hope the preceding chapters will have indicated the extent to which engine advancement has depended on the makers of components and accessories, and the high level of technology that these essential items require today. From small beginnings, the component industry has grown into a major contributor of great diversity, and it has done so principally by providing a good service to its customers, the engine manufacturers.

One message which should have come through in this history is that the successful specialist firm must have high competence in a number of areas. For a start, its management must be very familiar with the engine scene and hence with the latest trends, so that the company can be ready to meet new needs as they arise, or even to anticipate them. Maintaining a strong position in the world's marketplace necessitates a willingness to invest, not merely in equipment but also in *people*; the component industry relies heavily on its men and women, since it stands or falls by its exploitation of the many engineering and scientific skills.

These skills start in research and development, which can be orientated either towards advancing the product, in the technical sense, or towards ways of manufacturing it more cheaply or to higher standards of quality. Then there is the design function which has to convert the R & D lessons into hardware that not only does its job but can be produced at the right price. However ingeniously the designer solves his problems, he is wasting his time if his brainchild is beyond the capabilities of the production side or prices itself out of the market.

It follows that the component design office has to liaise closely with the production engineers, who themselves need to be able to adapt efficiently to changes in design, in materials or in the scale of manufacture. To facilitate production the larger companies frequently design and build their own special-purpose machinery and equipment; alternatively, they have them made to their requirements by firms specializing in such work.

Production departments also have to exert particularly strict quality control because of the influence of quality on the essential reliability. As engines become ever more complex, their overall reliability must fall unless that of the individual constituents is improved. On-line inspection of items during manufacture – much of it done automatically today – is playing an increasing role in component quality control. Assemblies, too, require to be inspected or tested on completion, which means that the right decisions have to be taken in the factory as to where 100 per cent checking is necessary or where batch or random checks will suffice.

A sermon preached more than once in the book concerns the necessity for real collaboration between the engine manufacturers and the component suppliers. Much of this collaboration is needed at the R & D stage, when something new is coming along on either side, but it extends also into the subsequent production-design phase and even into the manufacturing one – not least in respect of approval after completion. Nowadays, happily, most engine companies appreciate that an early approach by them on a new project is the best recipe for success. After all, the specialists come along quickly enough when they have something novel to offer!

Since we have been looking at the past and the present, what of the future? As I see it, the engine component industry will play an even bigger part in its maturity than it did in its youth and adolescence. This judgment is based primarily on the fact that the curve of progress for conventional power units has inevitably begun to flatten, so greater effort will be needed to achieve a forward step of given size. The component makers are certainly capable of providing that extra effort owing to their technical solidity and strong motivation. Should a completely new generation of engines appear, on the other hand, they have both the knowledge and the experience to solve the fresh set of problems such prime movers would generate.